MICRO-EVIDENCE FOR THE DYNAMICS OF INDUSTRIAL EVOLUTION: THE CASE OF THE MANUFACTURING INDUSTRY IN JAPAN AND KOREA

MICRO-EVIDENCE FOR THE DYNAMICS OF INDUSTRIAL EVOLUTION: THE CASE OF THE MANUFACTURING INDUSTRY IN JAPAN AND KOREA

ALMAS HESHMATI AND JEONG-DONG LEE
EDITORS

Nova Science Publishers, Inc.
New York

For permission to use material from this book please contact us:
Telephone 631-231-7269; Fax 631-231-8175
Web Site: http://www.novapublishers.com

LIBRARY OF CONGRESS CATALOGING-IN-PUBLICATION DATA

Lee, Jeong-Dong.
Micro-evidence for the dynamics of industrial evolution : the case of the manufacturing industry in Japan and Korea / Jeong-Dong Lee and Almas Heshmati.
p. cm.
ISBN 978-1-60456-388-7 (hardcover)
1. Manufacturing industries--Japan. 2. Manufacturing industries--Korea (South) I. Heshmati, Almas. II. Title.
HD9736.J32L44 2008
338.4'7670952--dc22

2007052751

Published by Nova Science Publishers, Inc. ✦ New York

CONTENTS

PREFACE

The creation of firms with new ideas, their growth, and the end of inefficient ones constitute the essence of "creative destruction" which has been the engine of capitalist societies. By observing the process of birth, growth, and death of firms, we can understand the overall picture of dynamics that an industry experiences over its life cycle.

This book contains nine chapters, all of which eight are contributed, and is divided into three parts. The first part shows some stylized facts of industrial dynamics in Korea and Japan. The second part deals with the issue of globalization and innovation from the perspective of industry dynamics. The last part investigates the implication of size on the Korean manufacturing industry. The common feature of all contributions is found in the approach used. Thus, the authors do not focus on the theoretical models but on numerical evidence.

Japan has taken the lead position in the level of technological and industrial performance as characterized by its competitively large companies and productive Small and Medium sized Enterprises (SMEs), while Korea has been known as one of the typical success stories of countries to catch up in the latter half of 20th century. During the catching-up process, Korea has become a so-called dual economy with a handful of globally competitive large companies and large number of weakly competitive SMEs. Therefore, the two countries can be considered good representative economies for comparative industrial analysis.

In: Micro-Evidence for the Dynamics of Industrial Evolution ISBN: 978-1-60456-388-7
Editors: A. Heshmati, J.-D. Lee

Chapter 1

CHANGE OF INDUSTRIAL STRUCTURE: IMPORTANCE OF MICRO EVIDENCE

Almas Heshmati and Jeong-Dong Lee
University of Kurdistan Hawler and Hawler Institute for Economic and Policy Research, Erbil, Federal Region of Kurdistan, Iraq
and
Seoul National University, Seoul, Republic of Korea

The creation of firms with new ideas, their growth, and the end of inefficient ones constitute the essence of "creative destruction", as termed by Schumpeter in the early 20th century, which has been the engine of capitalist societies. By observing the process of birth, growth, and death of firms, we can understand the overall picture of dynamics that an industry experiences over its life cycle (Carroll and Hannan, 2000). Productivity growth of an industry can be made by resource reallocation across firms from less to more productive ones, and the productivity enhancement of the individual firms as well as from advancement in technology. All these arguments signify the importance of micro firm-level data analysis to understand the dynamics of an economy. In addition, micro data deals directly with the issue of inherent heterogeneity of the economic agents and the interaction among them, which aggregated industry-level or national-level data cannot touch (Audretsch and Klepper, 2000). An analysis on the change of industrial structure through the window of micro data has relevance on important issues of industrial growth, such as innovative activity, globalization, regulation, institutional arrangement, and so on. This study, thus, is expected to provide not only managerial implications, but also policy directions on the changes of industrial structure.

This collection of applied research is one of the important outcomes of 'Asia Pacific Productivity Conference (APPC) 2006' which was held at the Seoul National University, Seoul, Korea in August 17-19, 2006. Researchers from many countries including Japan and Korea gathered and presented their research results and had discussions regarding the implications of their results. We agreed to make a separate publication to share our experience with other experts in this field. Thus, in this study, we focus on two countries, namely Japan and Korea, since the two countries are closely related in terms of trade. The two countries are competing on the one hand and cooperating on the other hand with similar institutional arrangements. Japan has taken the lead position in the level of technological and industrial performance as characterized by its competitively large companies and productive Small and Medium sized Enterprises (SMEs), while Korea has been known as one of the success stories of countries to catch up in the latter half of 20th century. During the catching-up process, Korea has become a so-called dual economy with a handful of globally competitive large companies and a large number of weakly competitive SMEs. Therefore, the two countries can be considered good representative economies for comparative industrial analysis.

The book contains nine chapters one of which is an introduction and the remaining eight are contributed chapters, and is divided into three parts. The first part shows some stylized facts of industrial dynamics in Korea and Japan. The second part deals with the issue of globalization and innovation from the perspective of industry dynamics. The last part investigates the implication of size on the Korean manufacturing industry. The common feature of all contributions is found in the approach used. Thus, the authors do not focus on the theoretical models but on numerical evidence. A short introduction of the individual chapters follows.

PART I: STYLIZED FACTS OF INDUSTRIAL DYNAMICS IN KOREA AND JAPAN

Lee, Oh, Heshmati and Baek (Chapter Two) analyze "the stylized facts of firm dynamics in Korean manufacturing in the 1990s and 2000s". They summarize the firm level evidence for the dynamics of the Korean manufacturing industry. The study is based on the most recent firm level data and covers the period when the Korean manufacturing underwent radical changes. Four main issues are looked at: (i) the overall picture of the productivity performance of the Korean manufacturing firms by size, sector and over time, (ii) the aggregate industry level productivity growth decomposed into sources attributed to entry and exit effects, (iii) the entry and exit pattern of the firms, and (iv) the survival of firms. By accounting for the financial crisis and its impacts on firms' behavior concerning change in their operation systems including financial structure, propensity to invest, and the strategies for competition, the authors expect that the empirical analysis might also catch the changes in the industry dynamics during the financial crisis. The result of the analysis of entry and exit show that with only sound metabolism embedded in the industry dynamics, the economy can bring a sustainable growth process. This study adds value to the existing literature on the competitiveness of the Korean manufacturing by focusing on the issue of productivity and survival, and paying special attention to the changes before and after the financial crisis and the review of the conventional management paradigm. They observe differences in firms' performance by size. In addition to the positive effects, the crisis exerted negative influence by the management style of LSE becoming more risk averse, hence lowering incentive to investment and focusing on the short-term productivity improvement throughout the restructuring period. The authors find the micro-level research useful and proposes continued research to accommodate international comparisons and sharing experience.

Fukao, Kim and Kwon (Chapter Three) in their study on "plant turnover and TFP dynamics in Japanese manufacturing" analyze the cause of the slowdown in Japan's TFP growth during the 1990s. Previous studies examining the issue at the macro- or industry-level, have found that the slowdown was primarily due to the stagnation of TFP growth in the manufacturing sector. The authors by using establishment level panel data covering the entire sector find that the reallocation of resources from less efficient to more efficient firms was very slow and limited. This "low metabolism" seems to be an important factor explaining the slowdown in Japan's TFP growth. In addition, the authors observe that the decline in the manufacturing was more than in the non-manufacturing sector. Persistency in productivity growth rate over time, the net entry productivity effect from exit and entry of firms are positive but low suggesting a weak creative destruction process. The decline in productivity growth was as a result of relocation of production abroad and within effect due to low start-ups and high close rate as main sources. These observations imply that an active turnover in plants is essential to boost Japan's aggregate productivity.

Ahn and Ito (Chapter Four) in their study on "the impact of internationalization on employment and productivity growth: a comparative analysis of Japanese and Korean micro-data" examine the economic impact of outbound FDI and trade by comparing the Korea and Japan by the means of regression analyses. The comparison of these two countries is motivated by the fact that: (i) Japan has a longer experience in outbound FDI and overseas production than Korea, (ii) Korea's outbound FDI is much more concentrated in China than Japan's which may explain the differences in the impact on the home economy, (iii) despite high global competitiveness of Korean firms in a number of areas, there still exists a significant gap between Korea and Japan in exporting firms productivity levels. The closer similarity in skills of workers in Korea and China suggest that Korean workers are much more strongly affected by competition from Chinese workers than the Japanese counterpart. This study reconfirms and provides a more nuance perspective on the previous finding of negative correlation between outbound FDI and growth in employment and productivity in manufacturing in Korea and Japan. However, the negative effect of outsourcing is more pronounced in the case of Japan with more advanced outsourcing activities than Korea. A focus on the link between the evolution of global production networks and their impacts on the regions economies might shed light on the differences in the countries responsiveness. Another intriguing finding is that the industry-level R&D intensity has a positive effect on the TFP growth in Korea but a negative effect in Japan. A possible explanation as to the difference might be the magnitude of spillover effects or the distance to the technological frontier where Korean firms easily imitate and absorb more advanced technologies. The authors propose more rigorous empirical studies based on micro-data to shed more light on the development process of firms' capabilities and firms' R&D strategies.

PART TWO: GLOBALIZATION, INNOVATION AND INDUSTRIAL DYNAMICS

Motohashi (Chapter Five) investigates the economic implications of "Globalization of Japanese Firms and its Impact on Domestic Economy". In the new era of globalization, large firms seek for attractive markets and cost advantage conditions by producing in low labor cost developing countries. Relocation of production site by the large firms' to overseas has a negative impact on SMEs production. SMEs do not have sufficient management resources for overseas production since their major customer has shifted their production sites. In this regard, Motohashi analyze the impact of globalization of large firms as a major player on domestic economy, with special emphasis on the business activities of SMEs by using domestic and overseas business surveys. The results suggest that despite a higher growth rate for globalized firms, there is no direct evidence of negative impact of globalization of production on SMEs in this study. The SMEs have gradually been adapted to the increasing globalization process by being more and more involved in global production activities or enhancing their competition by upgrading their manufacturing technology and product designs. This indicates presence of complementary relationship between domestic and overseas sales. Motohashi suggest similar analysis but based on manufacturing census data to reconfirm the accuracy of the survey based results.

Urata, Matsuura and Wei (Chapter Six) analyze the extent of the "international intrafirm transfer of management technology by Japanese multinational corporations" and its determinants by using firm level data. They examine the extent of technology transfer for a range of eight different job classifications. Two types of technology transfer involving MNCs can be identified: (i) technology transfer from parent firms of MNCs to their overseas affiliates, and (ii) technology transfer from overseas affiliates of MNCs to local firms. The former type is characterized as intrafirm and must take place before the latter as technology spillover is realized. Results show that top management has not been transferred to affiliates, but labor management requiring close local contact has. Statistical analysis shows that characteristics and strategies of parent firms, their affiliates and host countries affect the extent of technology transfer. The length of operation of the affiliates and the quality of labor are found to have positive impacts on technology transfer for the Asian affiliates. These findings indicate among others close local ties between sales and purchase, link between the length of operation of the affiliates and the host country level of industrialization, and that availability of high quality of labor promote technology transfer for the countries in Asia. The policy implications of the study include: provision of an FDI friendly environment by the countries, development of capable workers through education and training, development of competitive manufacturing sector to facilitate intrafirm technology transfer and also technology spillover to local firms. In order to deal with these policy challenges the authors suggest that, the government of the developing countries should formulate comprehensive development policies ranging from human resource development to FDI policies and implement them effectively.

Oh, Han and Heshmati (Chapter Seven) analyze "the relationship between innovation and performance of Korean manufacturing firms". This is in relation to the fact that the manufacturing industry in general and information technology related manufacturing and heavy industries in particular have been a principal contributor to growth of the Korean economy. The technology innovation has been recognized as a critical factor to the productivity growth of Korean firms. Therefore, the Korean government through various support and subsidy programs has induced innovation in manufacturing with the aim of improving R&D intensity of firms to enhance their productivity growth. Lack of data has been a limiting factor to establish the relationship between R&D inputs and productivity growth in the manufacturing sectors, hence causing difficulties to design public subsidy policies. Using combined innovation survey and financial data, the authors investigate the relationship between innovation and performance of Korean manufacturing firms. In particular, they identify determinants of innovation and investigate the links and causality relationships between innovation input and output, and between innovation output and labor productivity growth at the firm level. In their research, the authors use an up-to-date method that accounts for both simultaneity and selection bias. Results suggest that the actual linkage from R&D investment to the innovation output was not observed. This might have to do with the fact that the analysis is based on cross-sectional data and for obvious reasons it takes longer time to observe such a link. There is a positive association between the size of firms, the share of foreign ownership and number of patents registered with R&D investment and productivity growth. The results suggest that the current policy to enhance productivity of manufacturing sector need more complicated instruments other than just focusing on increased investment in R&D.

PART THREE: SIZE AND PERFORMANCE DISTRIBUTION

Baek and Lee (Chapter Eight) studied the "threshold effect on the growth of SMEs in the Korean manufacturing industry". It is motivated by the intentional retardation in growth of SMEs. The aim is to verify whether the policy support for SMEs retard the growth of SMEs found in survey data at the vicinity of 300 employees. This view is supported by the fact that billion dollars are given to SMEs but regulations are imposed on LSEs. The authors conduct a regression analysis with the database of all statutory audit firms to verify the existence of a threshold effect. The empirical results showed that the existence of a threshold effect in form of abnormal growth patterns cannot be verified in this analysis framework. The small sample size might cause difficulties in identifying the threshold. Most policy supports for SMEs were distributed to very small firms, and regulations on LSEs were imposed to very large firms with marginal effect. The authors suggest that more serious debates are required to reveal the efficacy of the dichotomous approach between SMEs and LSEs in industry policy.

Kim, Oh and Heshmati (Chapter Nine) analyze the "stochastic dominance structure of productivity distribution from firm dynamics perspective". This study is motivated by the competitive entry and exit process of firms which allows capital and labor to be reallocated from declining industries to growing industries which makes industry and the economy more dynamic. The authors aim at analyzing whether the effect on productivity from entry and exit process is positive or negative and comparing the distribution of productivity in the Korean manufacturing industry. The test for the first and second order stochastic dominance of total factor productivity level between exit groups and entry groups is conducted. The analysis is based on plant level panel data covering the Korean manufacturing during 1993 to 2003. Results show evidence of negative effect of productivity on entry and exit process which differs also over time, across different firm sizes classes below 300 employees, technology level and industrial sectors. Large size has no dominance in distribution of productivity growth. After the financial crisis in mid-1997 and 1998, the exit group with less than 50 employee second order dominated the entry group in 1999. Similar result was observed for plants with less than 20 employees in 2002. A positive effect would have come from the survival of higher productive firms and the death of lower productive firms. But continuous survival of lower productive firms and the death of higher productive firms in the industry lead to a negative net effect.

Micro firm-level data has become popular as the longitudinal data set becomes available from 1980s. The current volume might be the interim outcome of research in this field in two countries, namely Japan and Korea. For further development of the field in this region, we have to include more countries, especially Asian developing countries. The widened coverage of target country renders the researcher to understand the intricate relationship between the changes of global value chain and industry dynamics in the relevant countries. Accommodating case based study, which track the event in life cycle of firm and industry might be another route to develop further as indicated by Tybout (2000). One last issue to be considered can be found in the linkage with other key economic agenda, which takes importance as the global economy develops. The keywords to be considered , include Foreign Direct Investment (FDI), cluster development, outsourcing, regional growth disparity, to name a few. Co-editors together with all contributors hope that this is the start of a long mission and intensive research on industry dynamics and look forward for colleague members' contribution.

REFERENCES

Audretsch, D. B. and S. Klepper (eds.), 2000, *Innovation, Evolution of Industry and Economic Growth*, Cheltenham: Edward Elgar.

Carroll, G. R. and M. T. Hannan, *The Demography of Corporations and Industries, Princeton*, N. J.: Princeton University Press.

Tybout, J. R. (2000), Manufacturing Firms in Developing Countries: How Well Do They Do, and Why?, *Journal of Economic Literature*, 38 (1), 11-44.

In: Micro-Evidence for the Dynamics of Industrial Evolution ISBN: 978-1-60456-388-7
Editors: A. Heshmati, J.-D. Lee

Chapter 2

STYLIZED FACTS OF FIRM DYNAMICS IN THE KOREAN MANUFACTURING IN THE 1990S AND 2000S

Jeong-Dong Lee
Technology Management, Economics and Policy Program,
College of Engineering, Seoul National University, Seoul, Korea
Inha Oh *
Technology Management, Economics and Policy Program,
College of Engineering, Seoul National University, Seoul, Korea
Almas Heshmati
University of Kurdistan Hawler, and
Hawler Institute for Economic and Policy Research
Hawler, The Federal Region of Kurdistan, Iraq

ABSTRACT

The Korean economy suffered profoundly from the Asian financial crisis, and its rapid recovery in the following years has been studied extensively. One of the principal phenomena resulting from the crisis, and the subsequent unequal size-related recovery, was the exacerbation of the gap between large-sized enterprises (LSEs) and small and medium-sized enterprises (SMEs). In this paper, a comparative analysis was conducted by size classes and the sources of TFP growth were deconstucted into effects associated with the entry, exit, and survival of plants in the Korean manufacturing industry. Additional survival analysis was also conducted in order to assess the internal and external determinants for firm survival.

Keywords: Firm dynamics, TFP, survival analysis, SME, Korean manufacturing
JEL Classification Numbers: C33, C41, D24, L60, O30

---

* Corresponding author: Inha Oh, whistle3@snu.ac.kr, inha.oh@gmail.com

1. INTRODUCTION

The dynamic pattern of industry evolution can be characterized by the productivity performance of the constituent firms. In addition, firm entry and exit have also been shown to effect changes in the industrial landscape. Although the overall picture of firm dynamics can be captured via aggregated measures, including industry and/or national level productivity performance, the firm level evidence provides a more detailed description and also provides other insights into the sources of variations in industry performance. In this study, the firm level evidence for the dynamics of the Korean manufacturing industry is summarized, with special emphasis placed on the size of firms. This investigation is predicated on the most recent firm level dataset, and it encompasses the time period of the 1990s and 2000s, during which time the Korean manufacturing community underwent radical changes.

In a rapidly changing economy, small and medium-sized enterprises (SMEs) function as an accommodator of the flow of new products and process, and also nurture an environment characterized by innovation (Acs and Audretsch, 1990; Audretsch, 2002; Taymaz, 2002). Moreover, the increased share of employment and output of SMEs throughout the entirety of the Korean economy shows that SMEs constituted a vital constituent of industrial organizations[1]. The existing literature generally asserts the existence of significant productivity differences between SMEs and large-sized enterprises (LSEs), regardless of sector- and country-specific factors. However, the differential of productivity has been narrowed along with the level of economic development (Bartelsman et al., 2003; Snodgrass and Biggs, 1995).

The productivity gap between SMEs and LSEs has engendered an increasing number of comparative studies regarding the evolution and dynamics of firms. The majority of these studies have been influenced largely by two pieces of path-breaking literature: the passive learning model developed by Jovanovic (1982) and the active learning model proposed by Ericson and Pakes (1995). These models were predicated on the concept of 'creative destruction' ascribed to Joseph Schumpeter. The Schumpeterian model differs from 'standard theories of firm behavior' in that it recognizes firm heterogeneity, which is crucial to the development and creation of new processes, products, and markets.

In the passive learning model, a firm cannot predict its own potential profitability prior to market entry. Only after market entry can the firm assess its exit, expansion, or contraction options, by updating information the firm's revealed capability; for example, its relative position on the productivity distribution and its realized profit level. One of the principal implications of this model was that smaller and younger firms may evidence higher rates of growth. On the contrary, in the active learning model, a firm actively collects information both within and outside of the industry prior to market entry, and then decides the amount of investment necessary to maintain profitability under competition conditions. The potential profitability of the firm will change over time according to the stochastic response to its own investment and to other players' investment in the same market. The firm grows when it is successful and, conversely, diminishes when it is not.

The two models above agree on three stylized facts: (i) entrants start at a small scale compared to the incumbents; (ii) most new entrants evidence a lower probability of survival,

---

[1] For more detailed discussion, see the ensuing data section and Table 1.

and (iii) once an entrant's survival is certain, their growth rate is higher than the growth rate of the incumbents (Audretsch et al., 1999; Heshmati, 2001; Das, 1995). These stylized facts dictate that, competitive SMEs, if they survive, will function as an engine for productivity growth in the industry. As a result, the depression of SMEs in the Korean manufacturing sector, subsequent to the Asian financial crisis, have exerted a harmful effect on the future growth potential of the Korean economy.

This study attempts to assess the productivity performance of the Korean manufacturing industry, focusing principally on the differences between LSEs and SMEs. The overall productivity change is decomposed into entry, exit, and survival effects, in order to determine the implications with regard to the sources of industry-level performance changes. Survival analysis was also conducted in an attempt to analyze the factors contributing to firm survival in the Korean economy.

In the discussion of the results, we have lavished special attention on the changes in indicators occurring with the Asian financial crisis in the late 1990s. In their attempts to confront the financial crisis, the Korean manufacturing firms have been compelled to alter their operation systems, including their financial structures, propensity to invest, and competition strategies, etc., in order to fulfill the so-called 'global standard'. Thus, we expect that the empirical results discussed in this study may also bear import with regard to those changes occurring in the industry dynamics during the crisis period.

The paper is organized as follows; in section 2 we briefly introduce the different methodologies employed herein for the analyses of the productivity and survival of firms. The third section describes the data and the key variables utilized in the empirical analysis. The empirical results are discussed in the fourth section, and finally, the last section contains some concluding remarks.

2. METHODOLOGIES

Various methodologies were employed herein for the analysis of firm dynamics, including the measurement of total factor productivity (TFP), the observation in changes in productivity distribution over time, the decomposition of sources of TFP growth, and survival analysis. These analyses will delineate the pattern of productivity change per se, and will also describe the role of the reallocation process with entry, exit, expansion and contraction in the productivity changes at an industry level. In addition, our survival analysis enables us to assess the impact of the market environment and polices by determining the manner in which the market selection process and the institutional setting work in practice (Taymaz, 1997 Disney et al., 2003; Hahn, 2000; Baily et al., 1992).

2.1. Total Factor Productivity

The TFP of each plant or firm can be computed using the chained-multilateral index number approach developed by Good (1985), Good et al. (1997) and applied by Hahn (2000, 2004). It employs a separate reference point for each cross-section of observations, and then chain-links the reference points together over time. The reference point for a given time period was constructed as a hypothetical firm with (arithmetic mean) input shares and (geometric mean) input levels. Thus, the output, inputs, and productivity level of each firm in each year are assessed relative to the hypothetical firm at the base time period. This approach enables us to make transitive comparisons of productivity levels among all observations in a data panel.

Specifically, the productivity index for firm i at time t in our study, using the non-parametric approach outlined above, is measured as follows:

$$\ln TFP_{it} = (\ln Y_{it} - \overline{\ln Y_t}) + \sum_{\tau=2}^{t} (\overline{\ln Y_\tau} - \overline{\ln Y_{\tau-1}}) -$$
$$\left\{ \sum_{n=1}^{N} \frac{1}{2}(S_{nit} + \overline{S_{nt}})(\ln X_{nit} - \overline{\ln X_{nt}}) + \sum_{\tau=2}^{t} \sum_{n=1}^{N} \frac{1}{2}(\overline{S_{n\tau}} + \overline{S_{n\tau-1}})(\overline{\ln X_{n\tau}} - \overline{\ln X_{n\tau-1}}) \right\} \tag{1}$$

where Y, X, S, and TFP denote output, input, input share, TFP growth level, respectively, and symbols with a bar represent corresponding measures for hypothetical firms. The subscripts τ and n represent the reference time and inputs, respectively.

2.2. TFP growth Source Decomposition

Industry-level productivity growth can be decomposed into the internal resource reallocation effect due to the expansion and contraction of existing firms, and the external restructuring effect attributable to the entry and exit of firms. In this study, we applied the decomposition method developed by Griliches and Regev (1992) among the three alternative methods introduced by Disney et al. (2003), primarily because it renders an intuitive interpretation, and helps to avoid unnecessary measurement errors owing to the correlation between variables.

The output-weighted aggregated productivity in year t, P_t, can be defined as:

$$P_t = \sum_i \theta_{it} p_{it} \tag{2}$$

where θ_{it} was the output share of firm i in period t, P_t and p_{it} are output weighted and non-weighted productivity measures. Griliches and Regev (1992) decompose the change of the log transformed level of TFP *(ln TFP)* between t-k and t, denoted by ΔP) as follows:

$$\Delta P = \sum_{i \in S} \overline{\theta}_i \Delta p_{it} + \sum_{i \in S} \Delta\theta_{it}(\overline{p_i} - \overline{P}) + \sum_{i \in N} \theta_{it}(p_{it} - \overline{P}) - \sum_{i \in X} \theta_{i,t-k}(p_{i,t-k} - \overline{P}) \tag{3}$$

where S, N and X denote the set of firms that survived, entered, and exited, respectively, for the period from t-k to t. Here, the subscripts i and t indicate the firm and time, and the bar represents the average over the base and end years.

The first term on the right hand side in (3) indicates the 'within effect' as the contribution of productivity growth of the surviving firms to the aggregated industry-level productivity growth. The second term represents the 'share effect', which determines whether more productive firms acquire a larger market share. This effect measures the contribution of internal (within industry) restructuring among the surviving firms. The third and fourth terms indicate the contribution of the entry and exit of firms, respectively. If firms with a productivity higher than the industry average enter the market, then the entry effect becomes positive. If firms with lower productivity rates exit the market, the exit effect also becomes positive. The two entry and exit effects will represent the external (outside industry) restructuring effects. In this paper, we classify firms by size classes, as LSE and SME, such that we are able to disaggregate the above four effects in accordance with the transition across size classes, as follows:

- Within effect: LSE→LSE, LSE→SME, SME→LSE, SME→SME.
- Share effect: LSE→LSE, LSE→SME, SME→LSE, SME→SME.
- Entry effect: Enter as LSE, Enter as SME.
- Exit effect: Exit from LSE, Exit from SME.

2.3. Survival Analysis

Survival analysis was conducted herein in order to evaluate the determining factors for the survival of firms of different size classes. Specifically, the Cox proportional hazard model was employed in this study. The Cox proportional hazard model, introduced by Cox (1972), begins with the definition of the hazard rate of firms at time t_i as:

$$\lambda(t_i) = e^{-\beta' x_i} \lambda_0(t_i) \tag{4}$$

where the function λ_0 is the baseline hazard and x_i are covariates. A conditional operation in the Cox's partial likelihood allows for the estimation of β without requiring information of the baseline hazard, λ_0.

Traditionally, the set of covariates are generally assumed to be constant, which cannot be maintained in this case of firm survival. For example, GDP growth rate as one of the covariates affecting firm survival would change over time. Thus, the hazard function must be modified in order to accommodate the variable covariates as follows:

$$\lambda(t_i) = e^{-\{\beta' z_i + \gamma' x_i(t)\}} \lambda_0(t_i) \tag{5}$$

where z_i are time independent (or time invariant) covariates and $x_i(t)$ are time dependent covariates.

The covariates can each be identified as either external or internal. The external covariates are normally exogenous variables from the perspective of firms, for example, the GDP growth rate, which is the outcome of a stochastic process and thus is cannot be known a priori. The internal covariates are endogenous, and can be considered as outputs of the stochastic process generated by the observations under study. Considering the internal covariates as time-dependent requires restrictive assumptions of full understanding of the underlying stochastic process, such as the time lag structure (Fisher and Lin, 1999; Neumann, 1995; Dupuy and Mesbah, 2002 Duchesne and Lawless, 2000). In this paper, in order to maintain a minimum of required assumptions, only external covariates are assumed to be time-dependent and all internal covariates are regarded as time-independent.

The variables utilized as externals are macro variables, and are defined as:

- Industry growth rate (%) defined as annual growth rate in the number of firms included in the 3-digit industry level, and
- GDP growth rate (%) defined as the annual GDP growth rate.

The internal covariates are micro variables including:

- R&D dummy indicating whether the firm spent R&D expenditures
- Export dummy indicating whether the firm is engaged in the exporting activity
- Multi-plant dummy indicating whether the firm has multiple plants
- Number of employers as the number in the exit year, and
- Relative TFP level in the exit year as the TFP level of the firm compared to the 3-digit industry average TFP.

It should be noted that the stratification of hazard function was carried out at the 3-digit industry level.

3. Data and Variables

The data employed in this study was based on the unpublished plant-level data assembled from the Annual Report on Mining and Manufacturing Survey in Korea. The data covers all plants with five or more employees in 580 manufacturing industrial sectors at the KSIC (Korean Standard Industrial Classification) five-digit level. It is an unbalanced panel dataset with approximately 85,000 to 107,000 plants for each year from 1993 to 2002.

The unit of observation is the plant, not the firm. A plant is the most disaggregated unit of production and employment in the manufacturing industry, and plant turnover rather than firm turnover, is an important source of resource reallocation from a national economy standpoint. Although certain managerial decisions are made at the firm level, the real impact can be observed at the plant level, where the action takes place (Keller and Yeaple, 2003; Hakkala, 2004; Aw et al., 2002). Moreover, the single-plant samples constituted 93.3% of the whole sample and the implication of empirical analysis can be generalized to the firm case. Taking all this into consideration, we can find that a great deal of the empirical research thus far conducted on firm dynamics is based on plant-level data (Ahn et al., 2005; Ahn, 2001; Hahn, 2000).

In the survey data, the entry and exit of plants are identified on the basis of the appearance and disappearance of plants over time. During the study period, a total of 300,916 distinct plants were observed, and the total number of observations was 921,034. Plants with more (less) than 300 employees were designated LSEs (SMEs). The basic statistics regarding the number of plants, production volume, and employment are summarized in Table 1. From this table, we can see that the role of SMEs is becoming increasingly important with regard to production and employment in the Korean manufacturing industry.

For the TFP calculation, we employed the gross production of each firm as a measure of output, and capital, labor, and materials as input variables. The two-digit level industry price index was used to deflate the output variable. As a measure of capital stock, we utilize the average book value of capital stocks at the beginning and end of the year, deflated by the capital goods deflator[2] . As a measure of labor input, we utilized the number of workers, which includes paid employees (production and non-production workers), working proprietors, and unpaid family workers. In order to account for the labor quality factor, we reflect the quality index as the ratio of the non-production workers' and production workers' aggregate wage for each year. As a measure of intermediate input, we utilized the "major production costs" plus "other production costs" in our survey. The former encompasses costs associated with materials and parts, fuel, electricity, water, outsourced manufactured goods, and maintenance, and the latter includes outsourced services, such as advertising, transportation, communication, and insurance. The estimated intermediate input was deflated by the intermediate input price index. The cost shares were obtained as the respective input factors share of the total capital, labor, and materials costs.

We assume constant returns to scale when we compute TFP growth on the basis of a non-parametric approach. Therefore, the sum of factor input elasticities is assumed to be unity. Individual factor shares are assumed to be the average ones within the same plant size class[3] in the five-digit industry in a given year, thereby implying that the factor shares are allowed to vary across industries and size classes, as well as over time.

---

[2] In order to construct the capital stock series, we need the proxy of capital investment series and the level of initial stock, neither of which are available from the database. However, Baily et al. (1992) reported little difference between the productivity results with the book value of capital and those using constructed capital stock series, in a study of the US manufacturing industry.

[3] In calculating input factors, plants are grouped into three size classes according to the number of employees: 5-50, 51-300, and over 300.

Table 1. Numbers of plants, production and employment of the Korean manufacturing industry (1993-2002)

Year	Number of plants			Volume of production (Trillion Won in 2002 price)			Number of employees (Thousand)		
	Manufacturing industry	LSEs	SMEs	Manufacturing industry	LSEs	SMEs	Manufacturing industry	LSEs	SMEs
1993	85755	862	84893	285	139	146	2726	812	1914
(share; %)	(100)	(1.01)	(98.99)	(100)	(48.73)	(51.27)	(100)	(29.78)	(70.22)
1994	88422	842	87580	329	165	164	2771	816	1955
(share; %)	(100)	(0.95)	(99.05)	(100)	(50.06)	(49.94)	(100)	(29.45)	(70.55)
1995	93287	834	92453	375	187	188	2816	848	1968
(share; %)	(100)	(0.89)	(99.11)	(100)	(49.97)	(50.03)	(100)	(30.1)	(69.9)
1996	94050	830	93220	402	198	204	2780	850	1930
(share; %)	(100)	(0.88)	(99.12)	(100)	(49.16)	(50.84)	(100)	(30.56)	(69.44)
1997	89014	694	88320	397	187	210	2592	814	1778
(share; %)	(100)	(0.78)	(99.22)	(100)	(47.11)	(52.89)	(100)	(31.4)	(68.6)
1998	76357	575	75782	362	176	186	2122	566	1556
(share; %)	(100)	(0.75)	(99.25)	(100)	(48.56)	(51.44)	(100)	(26.66)	(73.34)
1999	88158	625	87533	422	203	219	2351	581	1770
(share; %)	(100)	(0.71)	(99.29)	(100)	(48.06)	(51.94)	(100)	(24.73)	(75.27)
2000	95361	661	94700	495	237	258	2514	609	1905
(share; %)	(100)	(0.69)	(99.31)	(100)	(47.85)	(52.15)	(100)	(24.21)	(75.79)
2001	103148	625	102523	521	240	281	2503	556	1947
(share; %)	(100)	(0.61)	(99.39)	(100)	(46.03)	(53.97)	(100)	(22.21)	(77.79)
2002	107482	615	106867	551	243	308	2541	527	2014
(share; %)	(100)	(0.57)	(99.43)	(100)	(44.08)	(55.92)	(100)	(20.75)	(79.25)
Overall	921034	7163	913871	4138	1973	2165	25716	6978	18738
(share; %)	(100)	(0.78)	(99.22)	(100)	(47.69)	(52.31)	(100)	(27.13)	(72.87)

4. Empirical Results

4.1. Development of the TFP by Size

Figure 1 indicates the output weighted average percentage growth rate of the TFP. Both size classes experienced negative growth rates during the crisis, but recovered rapidly in 1999 and 2000. We can see that the TFP growth rate of LSEs was higher than that of SMEs after the crisis.

The descriptive measures demonstrate that the TFP level has increased in the aggregate, and LSEs generally succeeded well in restructuring and downsizing to achieve higher productivity growth rate following the financial crisis. However, the productivity growth data of the SMEs is not indicative of similar recovery rates, thus suggesting that their restructuring in the market after the crisis has not been sufficient, and the TFP growth rate remains basically at a standstill. Another characteristic feature of SMEs' performance is higher dispersion among firms than SMEs', which is indicative of higher heterogeneity in that sector.

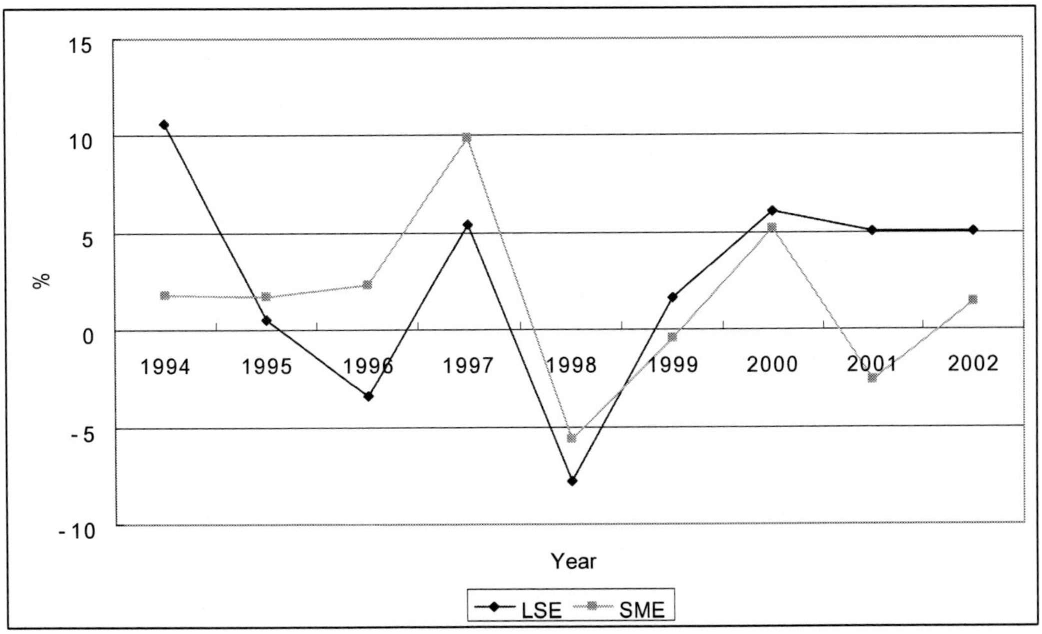

Figure 1. Output Weighted Average TFP Growth Rate (%)

4.2. Decomposition of the TFP Growth over Time

The result of the TFP growth decomposition over a 3-year span of time is reported in Table 2. The first sub-period of 1993-1995 represents the typical case for the pre-financial crisis, and the final sub-period of 2000-2002 denotes the case for post-financial crisis.

Firstly, we determined that the 'within effect' prior to the crisis (1993-1995) was positive and dominant in explaining the productivity growth of the manufacturing industry for both

LSEs and SMEs. However, the same 'within effect' becomes negative (-11.9%) in the case of SMEs following the crisis period (2000-2002). This means that, after the crisis, the TFP growth of surviving firms among the SMEs negatively influenced the industry-level TFP growth, whereas the survivors among LSEs positively affected the TFP growth rate. As can be witnessed in Figures 2 and 3, the gap in terms of profitability and R&D intensity by size classes has increased, particularly in the post-crisis period. LSEs, which accumulated profit and R&D capacity, have been successful with regard to innovation and internal restructuring for TFP growth. However, SMEs, which suffered severely during the crisis period, have no remaining innovative potential, and thus must manage to survive, to some degree with the help of government support.

Table 2. Source decomposition of TFP growth by period with three year time span.

Period			Sub-pediods			
			1993-1995	1997-1999	2000-2002	1993-2002
			(Pre-financial crisis period)	(Financial crisis period)	(Post-financial Crisis period)	(Entire Period)
Industry-wide TFP growth rate (%)			3.539	-3.155	2.036	1.925
			(100%)	(100%)	(100%)	(100%)
Composition of contributing effects (% to industry-wide growth rate)	Within (LSE)	LSE → LSE	90.888	42.622	77.982	31.025
		LSE → SME	4.097	0.663	6.075	5.858
	Within (SME)	SME → SME	36.829	42.696	-11.944	14.66
		SME → LSE	-1.165	4.656	-1.266	2.127
	Share (LSE)	LSE → LSE	-16.246	3.772	16.988	2.978
		LSE → SME	1.39	-2.065	3.128	-12.617
	Share (SME)	SME → SME	-15.459	-0.137	0.322	0.033
		SME → LSE	-0.943	0.335	0.044	-3.723
	Entry (Enter as LSE)		-7.671	5.716	10.162	13.805
	Entry (Enter as SME)		11.476	-17.965	23.487	19.629
	Exit (Exit as LSE)		5.349	-3.806	-0.225	19.308
	Exit (Exit as SME)		-8.544	23.513	-24.751	6.916

The 'Share effect' becomes positive after being negative across the financial crisis. As the share effect represents the intra-industry restructuring process, the positive sign demonstrates that the market mechanism that compensates for the productivity effort improves as the Korean economy moves through the crisis period (Jwa and Seo, 2000). However, the share effect of SMEs after the post-crisis period is negligible (0.32%), which is indicative of insufficient pressure toward market restructuring in the case of SMEs. Part of this story can be explained by the action of extensive public support programs designed to sustain the SME sector after the crisis.

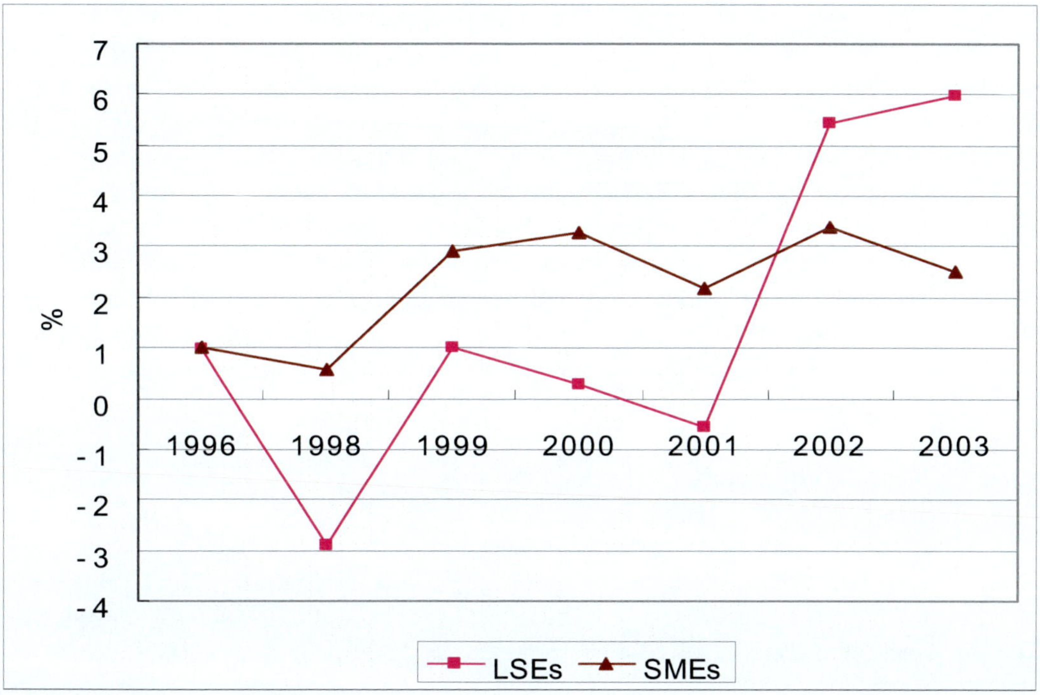

Figure 2. Development of profit ratio (Ordinary profits/Sales)for LSEs and SMEs (Bank of Korea).

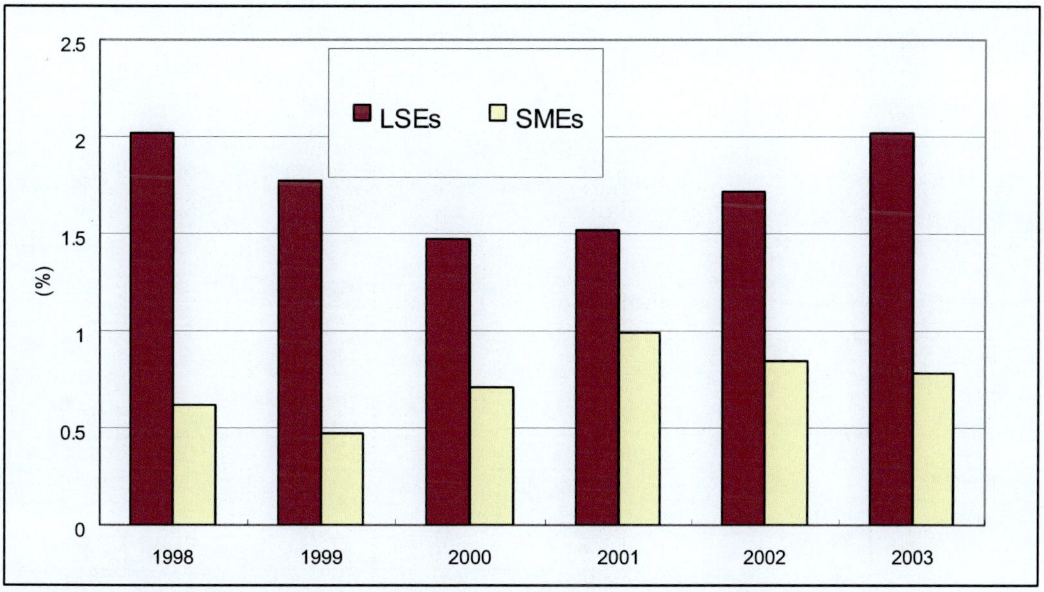

Figure 3. Development of R&D intensity according by size classes (Bank of Korea).

The 'entry effects' of the SMEs are positive both pre- and post-crisis, except for the turbulent period of the financial crisis, which implies the market entry of SMEs with productivity higher than the industry average. The same 'entry effect' of LSEs was negative for the pre-crisis period. The heavy intervention of government favoring LSEs prior to the crisis can explain why less productive LSEs entered the market. The IMF programs implemented during the financial crisis compelled the government to reduce the level of public intervention, particularly favoring LSEs. The positive sign of the entry effect of LSEs following the crisis might be the consequence of those changes in industrial policy.

The 'exit effect' of SMEs was consistently negative throughout the study period. As churning less productive SMEs is consistently a crucial source of industry-level productivity growth, the significantly negative impact of the exit effect in SMEs should be afforded due attention in the upgrading of productivity enhancements. The culprit behind this negative exit effect might be overly generous governmental support, allowing zombie firms to survive, as was described by Ahn et al. (2005) in a study of the Japanese manufacturing industry.

4.3. Cox Regression Results

Tables 3.a, 3.b, and 3.c provide the results of the Cox regressions of plant survival analysis for the case of the whole sample, LSEs, and SMEs, respectively. From Table 3.a, we can confirm that internal covariates, including R&D activity, exporting activity, etc. show the expected signs, with the exception of the productivity level. However, certain factors exert different impacts for LSEs and SMEs. For example, the positive impacts on survival owing to R&D and exporting activities are more prominent in the case of LSEs than in the case of SMEs. In addition, SMEs evidence a significant scale advantage, as the coefficient designating the number of employees is significantly negative, whereas the same effect is not significant in the case of LSEs. The positive sign of the TFP level for SMEs did not differ significantly from zero, thereby implying that the TFP level was not a crucial factor in terms of the survival of SMEs. This finding is consistent with the productivity decomposition results presented in the last section.

The overall message to be gleaned from this analysis is that both macro and micro variables are important to the survival of SMEs, whereas micro-variables were substantially more important to the survival of LSEs.

Table 3.a Cox regression results for all firms pooled.

External/Internal	Variable	Estimate	Error	Pr>Chisq	Hazard ratio
Externals	Sectorial Growth(%)	0.00783	0.00035	0.0001	1.008
	GDP Growth(%)	-0.02896	0.00076	0.0001	0.971
Internals	R&D	-0.54996	0.01126	0.0001	0.577
	Exporting	-0.57606	0.01117	0.0001	0.562
	Multi-plant	-0.10940	0.01248	0.0001	0.896
	# of employer	-0.00403	0.00020	0.0001	0.996
	TFP level	0.05994	0.00361	0.0001	1.062

Table 3.b Cox regression results for LSEs..

External/Internal	Variable	Estimate	Error	Pr>Chisq	Hazard ratio
Externals	Sectorial Growth(%)	-0.01086	0.01099	0.3230	0.989
	GDP Growth(%)	-0.02508	0.02667	0.3471	0.975
Internals	R&D	-0.70341	0.28923	0.0150	0.495
	Exporting	-1.28657	0.31725	0.0001	0.276
	Multi-plant	0.27058	0.29921	0.3658	1.311
	# of employer	0.00012	0.00011	0.2616	1.000
	TFP level	-0.95695	0.33584	0.0044	0.384

Table 3.c Cox Cox regression results for SMEs.

External/Internal	Variable	Estimate	Error	Pr>Chisq	Hazard ratio
Externals	Sectorial Growth(%)	0.00792	0.00035	0.0001	1.008
	GDP Growth(%)	-0.02912	0.00076	0.0001	0.971
Internals	R&D	-0.53318	0.01127	0.0001	0.587
	Exporting	-0.55545	0.01119	0.0001	0.574
	Multi-plant	-0.08803	0.01252	0.0001	0.916
	# of employer	-0.00682	0.00023	0.0001	0.993
	TFP level	0.06444	0.00367	0.0001	1.067

5. Concluding Remarks

An industry evolves principally with the entry of newly-born firms into the market with new and challenging ideas, technology, and unique management paradigms. Simultaneously, market forces compel the exit from the market of mismanaged firms with outdated technology and poor performance. With only sound metabolism embedded in the industry dynamics, the economy can robustly rejuvenate itself, and a sustainable growth process can be achieved.

This study addresses the dynamics of the Korean manufacturing industry based on plant level data collected between 1996 and 2003. The following are some of the stylized facts we discovered. Firstly, after the Asian financial crisis, the productivity growth rate of the SMEs was lower than that of the LSEs, which aggravates the performance gap between the two groups. Secondly, the restructuring process in the case of SMEs has not been satisfactory, as is shown by the significantly negative share effect and exit effect. Thirdly, productivity is not the principal determining factor with regard to the survival of SMEs.

All of the above findings imply that, even after the Asian financial crisis, the SMEs remain vulnerable to external shocks, and appear to be protected by a non-market-based support system. Another secondary source has shown that after the crisis, LSEs became more risk-averse, evidenced a lower investment propensity, and focused on short-term productivity improvements, principally by downsizing.

LSEs and SMEs differ considerably in their behaviors, and perform different functions in economic growth. On the basis of our study results, we conclude that the LSEs should pay more attention to upgrading their inner core competencies, such as R&D, and SMEs should be prepared to exploit more stringent market mechanisms, which would force them to take active steps to improve productivity.

REFERENCES

Acs, Z.J. and D.B. Audretsch, 1990, '*The Economics of Small Firms: A European Challenge*', Dordrecht: Kluwar.

Ahn, S., 2001, 'Firm dynamics and productivity growth: a review of micro evidence from OECD countries', OECD Economics Department Working Paper 297, OECD Economics Department.

Ahn, S., K. Fukao and H.U. Kwon, 2005, 'The internationalization and performance of Korean and Japanese firms: an empirical analysis based on micro-data', Discussion Paper 05008, Research Institute of Economy, Trade and Industry (RIETI).

Audretsch, D.B., 2002, 'The dynamic role of small firms: evidence from the U.S.', *Small Business Economics* 18, 13-40.

Audretsch, D.B., E. Santarelli and M. Vivarelli, 1999, 'Start-up size and industrial dynamics: some evidence from Italian manufacturing', *International Journal of Industrial Organization* 17, 965-983.

Aw, B.Y., 2002, 'Productivity dynamics of small and medium enterprises in Taiwan', *Small Business Economics* 18, 69-84.

Baily, M.N., C. Hulten and D. Campbell, 1992, 'Productivity dynamics in manufacturing plants', in Brookings Papers on Economic Activity: Microeconomics, Brookings Institute, Washington, D.C., 187-249.

Bartelsman, E., S. Scarpetta and F. Schivardi, 2003, 'Comparative analysis of firm demographics and survival: micro-level evidence for the OECD countries'. OECD Economics Department Working Papers 348, OECD Economics Department.

Cox, D.R., 1972, 'Regression models and life tables (with discussion)', *Journal of the Royal Statistical Society Series B* 34, 187-220.

Das, S., 1995, 'Size, age, and firm growth in an infant industry: the computer hardware industry in India', *International Journal of Industrial Organization* 13, 111-126.

Disney, R., J. Haskel and Y. Heden, 2003, 'Restructuring and productivity growth in UK manufacturing', *The Economic Journal* 113, 666-694.

Duchesne, T. and J. Lawless, 2000, 'Alternative time scales and failure time models', *Lifetime Data Analysis* 6, 157-179.

Dupuy, J.F. and M, Mesbah, 2002, 'Joint modeling of event time and nonignorable missing longitudinal data', *Lifetime Data Analysis* 8, 99-115.

Ericson, R. and A. Pakes, 1995, 'Markov perfect industry dynamics: a framework empirical analysis', *Review of Economic Studies* 62(1), 53-82.

Fisher, L.D. and D.Y. Lin, 1999, 'Time-dependent covariates in the Cox proportional-hazards regression model', *Annual Review of Public Health* 20, 145-157.

Good, D.H., 1985, '*The Effect of Deregulation on the Productive Efficiency and Cost Structure of the Airline Industry*', Ph.D. dissertation, University of Pennsylvania.

Good, D.H., M.I. Nadiri and R.C. Sickles, 1997, 'Index Number and Factor Demand Approaches to the Estimation of Productivity', in H. Pesaran and P. Schmidt (eds.), *Handbook of Applied Econometrics: Microeconometrics, Vol II*, Blackwell Publishers.

Griliches, Z. and H. Regev, 1992, 'Productivity and firm turnover in Israeli industry 1979-88', NBER Working Paper 4059.

Hahn, C.H., 2000, 'Entry, exit and aggregate productivity growth: micro evidence on Korean manufacturing', OECD Economics Department Working Papers 272, OECD Economics Department.

Hahn, C.H., 2004, 'Exporting and Performance of Plants: Evidence from Korean Manufacturing', NBER Working Paper 10208.

Hakkala, K., 2004, 'Corporate restructuring and labor productivity growth', IUI Working Paper 619, The Research Institute of Industrial Economics.

Heshmati, A., 2001, 'On the growth of micro and small firms: evidence from Sweden', *Small Business Economics* 17(3), 213-228.

Jovanovic, B., 1982, 'Selection and the Evolution of Industry', *Econometrica* 50(3), 649-670.

Jwa, S-H. and J-H. Seo, 2000, 'Financial crisis and perspectives on Korean economic development', in Choi, J. (ed.), *Asian Financial Crisis: Financial, Structural and International Dimensions,* Amsterdam: JAI.

Keller, W. and S.R. Yeaple, 2003, 'Multinational enterprises international trade, and productivity growth: Firm-level evidence from the United States', NBER Working Paper 9504.

Neumann, G.R., 1995, 'Search models and duration data', Labor and Demography 9511001, Economics Working Paper Archive at WUSTL.

Salim, R.A. and K.P. Kalirajan, 1999, 'Sources of output growth in Bangladesh food processing industries: a decomposition analysis', *The Developing Economies* 37(3), 355-374.

Snodgrass, D.R. and T. Biggs, 1995, '*Industrialization and the small firm patterns and policies*', San Francisco: International Center for Economic Growth.

Taymaz, E., 1997, '*Small and medium-sized industry in Turkey'*, Ankara: State Institute of Statistics.

Taymaz, E., 2002, 'Were small firms really less productive? An analysis of productivity differentials and firm dynamics', Paper presented at International Workshop on The post-entry performance of firms: technology, growth and survival, University of Bologna, 22-23 November 2002.

In: Micro-Evidence for the Dynamics of Industrial Evolution ISBN: 978-1-60456-388-7
Editors: A. Heshmati, J.-D. Lee

Chapter 3

PLANT TURNOVER AND TFP DYNAMICS IN JAPANESE MANUFACTURING

Kyoji Fukao*
Hitotsubashi University and RIETI, Tokyo, Japan
Young Gak Kim
Hitotsubashi University, Tokyo, Japan
Hyeog Ug Kwon
Nihon University, Tokyo, Japan

ABSTRACT

This study analyzes the cause of the slowdown in Japan's TFP growth during the 1990s. Many preceding studies, examining the issue at the macro- or industry-level, have found that the slowdown was primarily due to the stagnation in TFP growth in the manufacturing sector. Using establishment level panel data covering the entire sector, we investigate the causes of the TFP slowdown and find that the reallocation of resources from less efficient to more efficient firms was very slow and limited. This "low metabolism" seems to be an important reason for the slowdown in Japan's TFP growth.

JEL Classification Numbers: O4, O53

1. INTRODUCTION

In the 1990s, Japan experienced a significant slowdown in the growth of total factor productivity (TFP) – a trend that stands in stark contrast with many other advanced countries. A number of empirical studies have examined the causes of this slowdown in TFP growth by

* Correspondence: Kyoji Fukao, Institute of Economic Research, Hitotsubashi University, Naka 2-1, Kunitachi, Tokyo 186 JAPAN. Tel.: +81-42-580-8359, Fax: +81-42-580 -8333, e-mail: k.fukao@srv.cc.hit-u.ac.jp

focusing on the macro level.[1] Although these studies have thrown light on some aspects of Japan's economic malaise, few studies showed how much of the TFP slowdown at the macro level can be explained by the diseases they diagnosed.

In the case of the United States, Canada, and the EU countries, there has been a dramatic increase in the number of productivity studies using longitudinal micro-level data sets. In the EU KLEMS project, for example, Eric J. Bartelsman and other scholars try to link their results on productivity at the micro-level with sector- and macro-level results on productivity. Japan, however, probably because of the lack of appropriate data sets, has been left relatively behind in such studies, which link micro-level productivity analyses with sector- or macro-level productivity analyses.

In order to fill this gap, the research project "Study on Industry-Level and Firm-Level Productivity in Japan" at the RIETI (Research Institute of Economy, Trade and Industry) has been compiling a sector- and micro-level database with, in the case of sector-level data, covers the whole economy or at least, in the case of micro-level data, a substantial part of the Japanese economy. The sector-level data set is called the Japan Industrial Productivity (JIP) Database.[2]

Using the JIP 2006 Database and establishment-level data of the *Census of Manufactures* compiled by this project, this paper investigates the causes of the stagnation of productivity in Japan.

In this paper we first show that productivity slowdown mainly occurred in Japan's manufacturing sector. Next, we decompose the aggregate productivity growth of Japan's manufacturing sector into a within effect, a reallocation effect, and a net entry effect. Although Nishimura, Nakajima and Kiyota (2005), Fukao, Kwon (2004, 2006) and Ahn, Fukao and Kwon (2004) have already conducted productivity decompositions, using the firm-level data of the Ministry of Economy, International Trade and Industry's *Kigyo Katsudo Kihon Chosa* (*Basic Survey on Business Activities by Enterprises*), these studies do not allow an examination of the causes of Japan's productivity slowdown because the *Basic Survey* only covers the period after 1992.

In this paper, we assess changes in productivity dynamics at the sector- and the micro-level from the 1980s to 2003 using the JIP 2006 Database and the establishment-level data of the *Census of Manufactures*. To the best of our knowledge, this paper is the first study that compares Japanese micro-level productivity dynamics in the 1990s with those of the 1980s.

Our analysis suggests that the decline in aggregate productivity growth was not due to a decline in the reallocation effect but due to (a) a slowdown in TFP growth within establishments; and (b) a worsening negative exit effect, meaning that closures of productive establishments were more frequent than closures of unproductive establishments.

Finally, we examine why the negative exit effects have worsened and productivity growth within each establishment has drastically declined since the collapse of the bubble economy. We found some evidence supporting the hollowing-out hypothesis, namely, that productive establishments seem to be closed down and production relocated abroad. We also

[1] See Fukao and Kwon (2006) for overview of empirical studies on Japan's TFP growth rate of the 1990s at the macro- and the sector-level.

[2] The original version of the JIP Database (ESRI/Hi-Stat JIP Database 2003) was compiled in collaboration between ESRI (Economic and Social Research Institute, Cabinet Office, Government of Japan) as part of its research project on "Japan's Potential Growth" and Hitotsubashi University as part of its Hi-Stat project (A 21st-Century COE Program, Research Unit for Statistical Analysis in the Social Sciences).

found that the slowdown in productivity growth within existing establishments is partly caused by the low start-up rate and the high closure rate during the lost decade.

The remainder of this paper is organized as follows: In the next section, using the JIP 2006 database, we present an overview of Japan's TFP growth from 1970 to 2002. In Section 3, we conduct a decomposition of TFP growth in the manufacturing sector and compare our results with preceding studies on other developed economies. In Section 4, we examine TFP dynamics and entry and exit at the establishment level. Finally, in Section 5, we summarize our results and discuss the policy implications of our findings.

2. Evidence of Japan's Macro and Sectoral TFP Growth from 1970 to 2002

Since the early 1990s, Japan has suffered a large decline in aggregate productivity growth when compared to the relatively rapid growth in the 1980s (Hayashi and Prescott 2001, Fukao et al. 2004).[3] Sector-level analysis shows that the decrease in productivity growth can be seen across the board sector (Fukao et al. 2004, Fukao and Kwon 2006). In addition, previous studies have shown that the slowdown in TFP growth was more severe in the manufacturing sector than in the non-manufacturing sector. From an international perspective, Japan's productivity growth over the past decade or so has been extraordinarily low, and in contrast with Japan, many countries enjoyed significant improvements in productivity as a result of the ICT revolution in the 1990s.

In this section, we provide a detailed examination of the reasons for the slowdown in Japan's productivity growth during the 1990s using growth accounting. We will use aggregate and sectoral data from the JIP 2006 database developed by Fukao et al. (2006).[4]

Table 2.1 summarizes the result of the macro-level growth accounting. In our growth accounting exercise, we divided the period 1970-2002 into five-year intervals and into three long subperiods. We included the three long subperiods because growth accounting for shorter periods tends to be more strongly affected by cyclical variations in productivity growth. According to our calculations based on data from the JIP database, real GDP growth, which we calculated using a Laspeyres chain index, declined from 4.41 percent in 1980-1990 to 1.10 percent in 1990-2002.[5] This decline of 3.31 percentage-points can be decomposed into the following factors:

[3] It should be noted, however, that there are also several studies that came to the conclusion that the slowdown in the TFP growth in the 1990s was not that large. See, e.g., Jorgenson and Motohashi (2004) and Kawamoto (2004).

[4] The JIP 2006 Database was compiled as part of a RIETI research project. The detailed results of this project are reported in Fukao et al. (2006). The database contains annual information on 108 sectors, including 56 non-manufacturing sectors, from 1970 to 2002. These sectors cover the whole Japanese economy. The database includes detailed information on factor inputs, annual nominal and real input-output tables, as well as some additional statistics, such as capacity utilization rates, Japan's international trade by trade partner, inward and outward FDI, etc., at the detailed sectoral level. An Excel file version of the JIP2006 Database is available on RIETI's web site.

[5] As Fukao and Kwon (2006) indicate, the fixed-weighted Laspeyres price index for ICT products suffers from bias because the outputs share of ICT products, which registered faster price declines, increased.

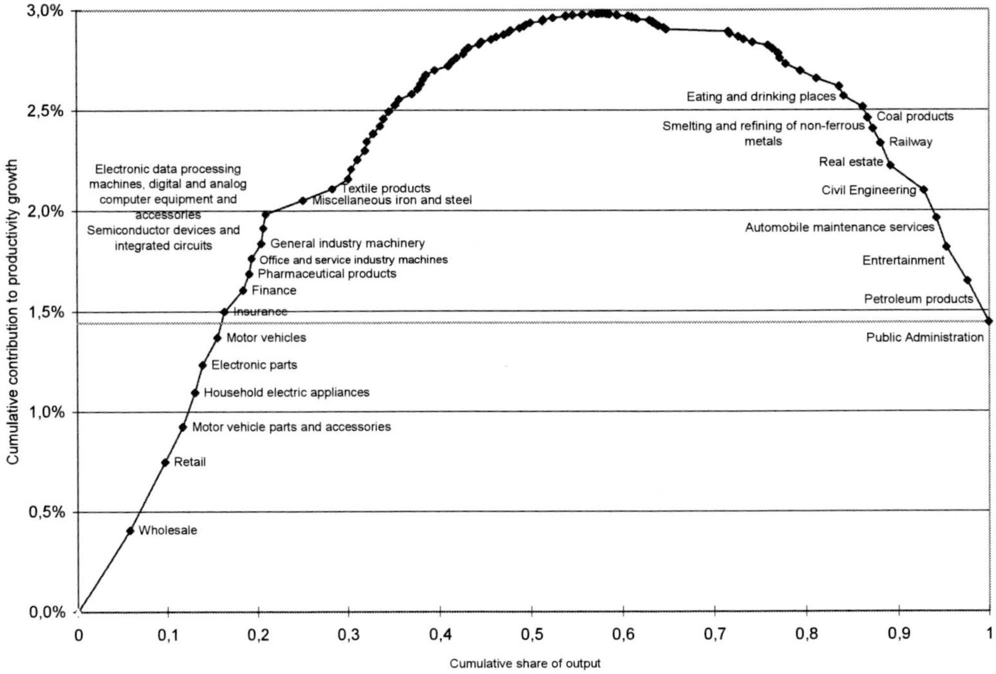

Figure 2.1.Cumulative Contribution of Industries to TFP Growth (1970-1980, Macro).

a decline in man-hour growth of 0.86 percentage points;
a decline in labor quality growth of 0.08 percentage points;
a decline in capital stock growth of 0.82 percentage points;
a decline in capital quality growth of 0.40 percentage points;
a decline in TFP growth of 1.17 percentage points.

As this list shows, all factors contributed to the decline in economic growth during Japan's lost decade. The decline in TFP growth rate is smaller than the results of Hayashi and Prescott (2002) and Yoshikawa and Matsumoto (2001), suggesting that the TFP growth rate at the macro-level declined by more than 2 percentage points from the 1980s to the 1990s. Our results also show that the most important factor underlying the stagnation during the 1990s is the decline in TFP growth. But what is also remarkable is the large drop in the growth contribution of capital. In fact, the slowdown in capital deepening – i.e., the contribution of the change in capital stock and capital quality taken together – was even more severe than the decline in TFP growth. These results indicate that the stagnation of Japan's economy has been caused by both supply-side and demand-side factors.

Table 2.2 compares the growth accounting results for the manufacturing and the non-manufacturing sectors. The results indicate that real output growth and man-hour growth slowed much more significantly in the manufacturing than in the non-manufacturing sector. On the other hand, the growth contribution of capital in the non-manufacturing sector, which had been substantial in the 1970s and 1980s, fell markedly during the 1990s. Finally, like many previous studies, we found that the slowdown in TFP growth was more severe in the manufacturing than in the non-manufacturing sector.

Looking at the estimates for the five-year intervals, we find that all growth indicators – real GDP, man-hours, capital stock, and TFP – saw a remarkable drop immediately after the collapse of the bubble economy. TFP growth, capital deepening, and labor supply all saw a much more pronounced deceleration than in previous business cycles. Following the collapse of bubble economy, Japan's economy failed to rebound quickly and TFP growth and capital deepening stopped.

Considerable differences in productivity growth across industries can be observed and it is quite possible that a few industries account for most of the productivity growth at the macro level. We therefore examined the contribution of individual sectors to macro-level productivity growth and the results are illustrated in Figures 2.1, 2.2, and 2.3. Figure 2.3 shows that during the "lost decade" of the 1990s, the wholesale sector accounts for all the productivity growth in the economy as a whole. In contrast, during the preceding decade, productivity growth in the economy as a whole was supported by eleven sectors. The aggregate productivity acceleration in 1980s was supported by eleven sectors, led by the wholesale, finance, retail, and civil engineering sectors due their relative size (Figure 2.2). In the 1970s, six sectors accounted for 95 percent of total productivity growth. These sectors included the motor vehicle parts and accessories, electronic parts, household electric appliances, and motor vehicles sectors, which registered the highest rates of productivity growth during this period.

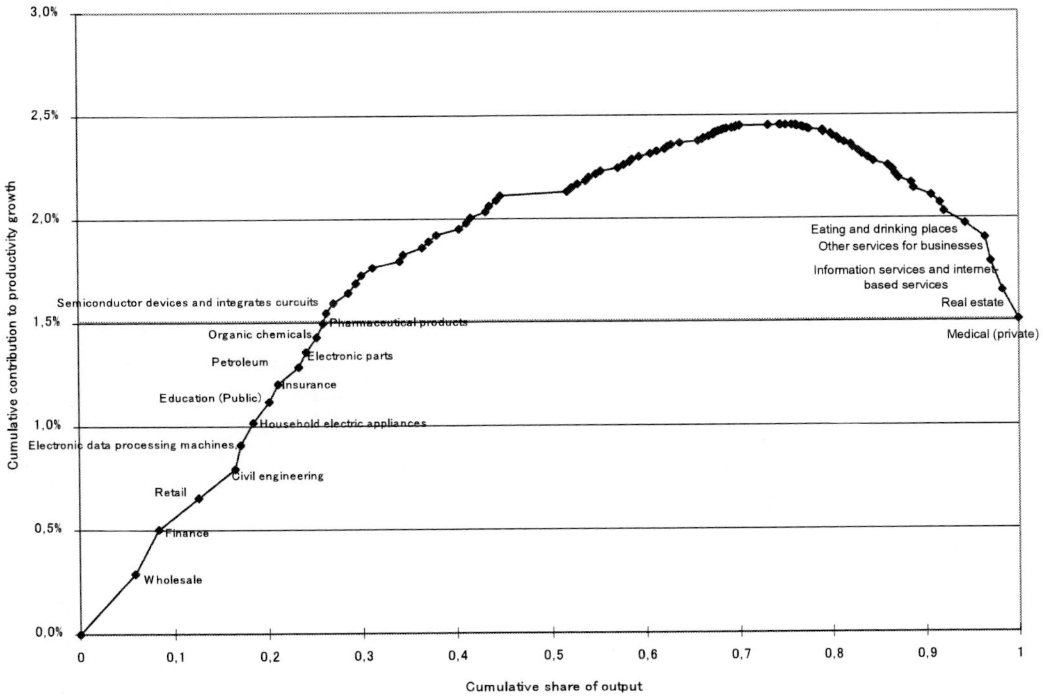

Figure 2.2. Cumulative Contribution of Industries to TFP Growth (1980-1990, Macro).

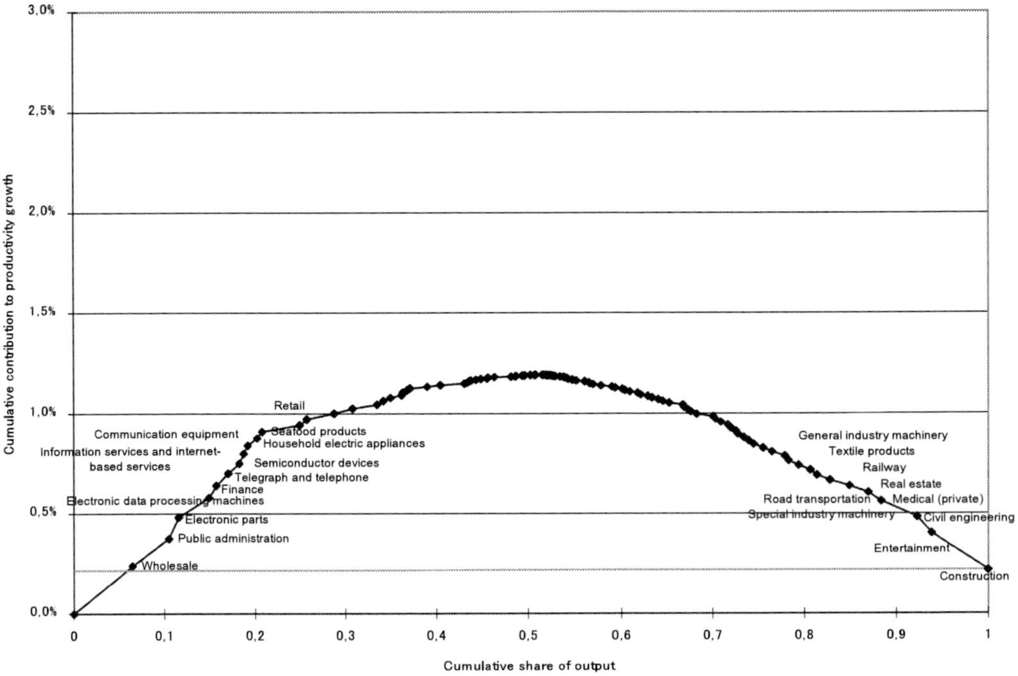

Figure 2.3. Cumulative Contribution of Industries to TFP Growth (1990-2002, Macro).

The rank of leading industries has remained fairly stable over the decades. The top five industries in the 1970s were: wholesale, retail, motor vehicle parts and accessories, electronic parts, and household electric appliances. In the 1980s, the order was as follows: wholesale, finance, retail, civil engineering, and electronic data processing machines. Finally, in the 1990s, the list was again topped by wholesale, followed by public administration, electronic parts, finance, and electronic data processing machines.

Comparing the top ten contributing sectors in Japan and the United States, we find that these are fairly similar.[6] Two sectors, wholesale and electronic parts, are among the top ten in the two countries in all periods. Overall, the two major industries making the greatest contribution to aggregate productivity growth both in Japan and the United States and throughout the entire period were the commerce and the electrical machinery/electronics sector.

During the lost decade, all industries experienced a decline of productivity growth, and sectors with strong productivity growth could not maintain their acceleration of productivity. There was good news such as five of the top ten contributing industries to productivity growth hailed from the service sector, such as "telegraph and telephone," "finance" and "information services and internet-based services." The above results indicate that for Japan's economy to once again achieve sustained rates of growth, the emergence of new growth sectors and an acceleration of productivity growth in the service sector are indispensable.

[6] See Farrell, Baily and Remes (2005) for case of United States.

3. Decomposition of Total Factor and Labor Productivity Growth in the Manufacturing Sector by Industry

As Baily, Hulten and Campbell (1992) and Foster, Haltiwanger and Krizan (1998) have shown in their productivity decomposition analyses, the start-up of productive establishments and the closure of unproductive establishments substantially contributed to TFP growth in the United States. Figure 3.1 shows that, in the 1980s, the start-up rate (the number of newly set up establishments divided by the number of all establishments) and the closure rate in Japan were only about half of the corresponding values for the U.S. Moreover, the gap widened in the 1990s as the start-up rate in Japan's manufacturing sector declined to only about 2%. This factor is likely to have contributed to the slowdown in TFP growth in Japan's manufacturing sector. We examine this crucial issue in this section.

We use a longitudinal database on Japanese manufacturing establishments for the period 1981-2003. The establishment-level data are taken from the *Kogyo Tokei Chosa* (*Census of Manufactures*), which is conducted annually by the Ministry of Economy, Trade and Industry (METI).[7] We created this longitudinal database making use of work of the "Quantitative Analysis Database Group" at RIETI on the linkage of establishment identification numbers over time.[8]

The census covers all Japanese manufacturing establishments except those belonging to the government as well as head offices not directly engaged in the manufacturing, processing or repair of industrial products. The census covers all establishments in years ending with 0, 3, 5 and 8 of the Western calendar year. In the other years, the Census covers establishments with four or more employees. In the case of censuses covering all establishments, data on establishments with less than four employees until a few years ago used to be managed and stored by prefectural governments. Most of these data appear to already have been discarded and therefore were not available to us.[9] For this reason, we had to confine our analysis to establishments with four or more employees. It is important to note that in this study, our data on establishments that "closed" include establishments which shrank to a size of less than four employees or changed their main business from the manufacturing sector to another sector.

There are 9,049,011 establishment observations for the period of 1981-2003. Out of these 9 million observations, we can calculate the labor productivity (real value added per man-hour) for about 8,852,575 observations and TFP for about 3,485,030 observations. Many small establishments did not provide information regarding key variables, such as capital stock, which are indispensable for our calculation of TFP. We treated establishments that failed to provide such information in one year as ongoing establishments, not as closed establishments, but we did not include these establishments in our calculation of industry-level TFP growth. This, however, means that our decomposition analysis of TFP growth at the industry level might be biased as a large number of small firms are not included. In order

[7] The compilation of the microdata of the *Census of Manufactures* was conducted as part of the RIETI project "Study on Industry- and Firm-Level Productivity in Japan."

[8] This group at RIETI includes Kazunari Shinbo, Mutsuharu Takahashi, Hyeog Ug Kwon, Toshiyuki Matsuura and others. For details on this linkage process, see Shinbo, Takahashi and Omori (2005).

[9] For more on this issue, see Shimizu and Miyagawa (2003).

to examine whether our TFP growth estimates are biased and to correct for this problem, we also decompose labor productivity by industry.

We divide the manufacturing establishments into 48 industries and calculate the relative TFP and labor productivity of each establishment in relation to the industry average. Following Good, Nadiri, and Sickles (1997) and Aw, Chen, and Roberts (1997), we define the TFP level of establishment f in year t in a certain industry in comparison with the TFP level of a hypothetical representative establishment in year 0 in that industry by

$$\ln TFP_{f,t} = (\ln Q_{f,t} - \overline{\ln Q_t}) - \sum_{i=1}^{n} \frac{1}{2}(S_{i,f,t} + \overline{S_{i,t}})(\ln X_{i,f,t} - \overline{\ln X_{i,t}}) + \sum_{s=1}^{t} (\overline{\ln Q_s} - \overline{\ln Q_{s-1}}) - \sum_{s=1}^{t} \sum_{i=1}^{n} \frac{1}{2}(\overline{S_{i,s}} + \overline{S_{i,s-1}})(\overline{\ln X_{i,s}} - \overline{\ln X_{i,s-1}})] \quad (3.1)$$

where $Q_{f,t}$, $S_{i,f,t}$, and $X_{i,f,t}$ denote the gross output of establishment f in year t, the cost share of factor i for establishment f in year t, and establishment f's input of factor i in year t, respectively. Variables with an upper bar denote the industry average of that variable. We use 1981 as the base year 0. We assume constant returns to scale. As factor inputs, we take account of capital, labor and real intermediate inputs.

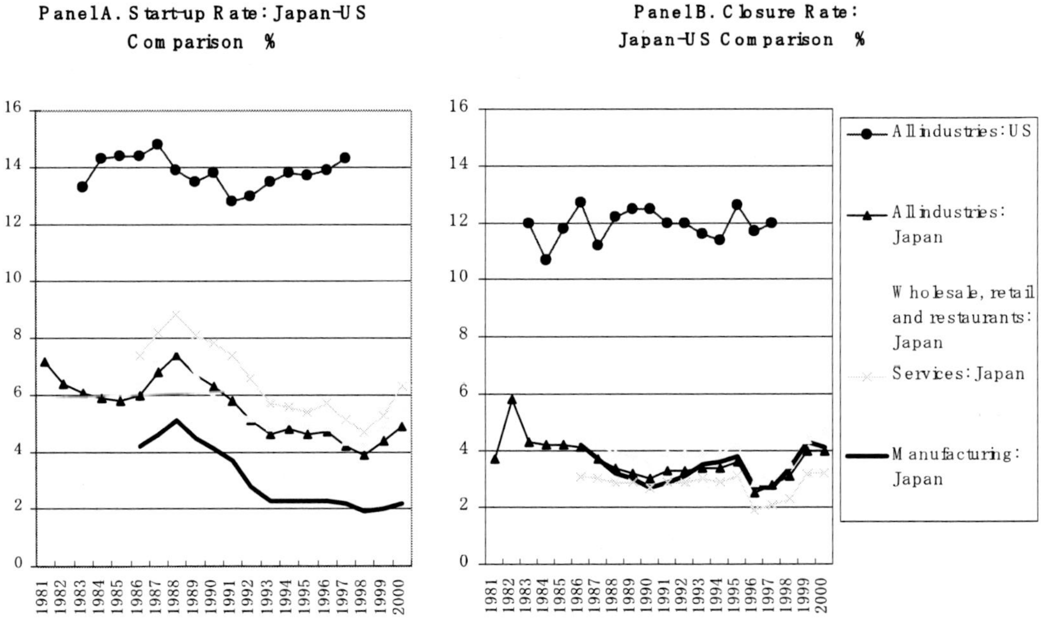

Sources: New Business Creation Subcommittee, New Growth Policy Committee, Industrial Structure Council (2002). The original data are taken from Small Business Administration, U.S. Government (1998), Small and Medium Enterprise Agency, Ministry of Industry, Trade and Industry, Japanese Government (2001), and Study Group on "Industry Hollowing-out" and Tariff Policy, Ministry of Finance, Japanese Government.

Figure 3.1. Start-up and Closure Rate of Establishments: Japan-U.S. Comparison.

We define the representative establishment for each industry as a hypothetical establishment whose gross output as well as input and cost share of all production factors are identical with the industry average. The first two terms on the right hand side of equation (3.1) denote the gap between establishment *f*'s TFP level in year *t* and the representative establishment's TFP level in that year. The third and fourth terms denote the gap between the representative establishment's TFP level in year *t* and the representative establishment's TFP level in year 0. Therefore, $lnTFP_{f,t}$ in equation (3.1) denotes the gap between establishment *f*'s TFP level in year *t* and the representative establishment's TFP level in year 0.

In a similar way, we define the labor productivity level of establishment *f* in year *t* in a certain industry in comparison with the labor productivity level of a hypothetical representative establishment in year 0 in that industry by

$$\ln LP_{f,t} = (\ln Y_{f,t} - \overline{\ln Y_t}) - (\ln L_{f,t} - \overline{\ln L_t}) + \sum_{s=1}^{t} (\overline{\ln Y_s} - \overline{\ln Y_{s-1}}) - \sum_{s=1}^{t} (\overline{\ln L_s} - \overline{\ln L_{s-1}}) \tag{3.2}$$

where $Y_{f,t}$, and $L_{i,f,t}$ denote the real value added (real gross output minus real intermediate input) of establishment *f* in year *t* and establishment *f*'s labor input in year *t*, respectively.

For details on the definition of and data source for each variable, please see Appendix A. Because of data limitations, we cannot take account of the change in labor quality in our productivity analysis. It is probably for this reason that we arrive at a higher TFP growth estimate than the industry-level result based on the JIP 2006 in the previous section. We also assume that the working hours at each establishment are equal to the industry average.

Adopting the methodology used by Baily, Hulten and Campbell (1992) and Foster, Haltiwanger and Krizan (1998), we define the industry-level TFP of a certain industry in year *t* by

$$\ln TFP_t = \sum_{f \in F} \theta_{f,t} \ln TFP_{f,t} \tag{3.3}$$

where $\theta_{f,t}$ denotes establishment *f*'s sales share in year *t* in that industry. *F* is the set of all the establishments existed at least either in year $t\text{-}\tau$ or in year *t* in this industry. Then, as Foster, Haltiwanger and Krizan (1998) showed, we can decompose the manufacturing sector's TFP growth from year $t\text{-}\tau$ to year *t*, $lnTFP_t - lnTFP_{t-\tau}$, into the following five factors.

Within effect: $\sum_{f \in S} \theta_{f,t-\tau} \Delta \ln TFP_{f,t}$

Between effect: $\sum_{f \in S} \Delta\theta_{f,t} (\ln TFP_{f,t-\tau} - \overline{\ln TFP_{t-\tau}})$

Covariance effect: $\sum_{f \in S} \Delta\theta_{f,t} \Delta \ln TFP_{f,t}$

Entry effect: $\sum_{f \in N} \theta_{f,t} (\ln TFP_{f,t} - \overline{\ln TFP_{t-\tau}})$

Exit effect: $\sum_{f \in X} \theta_{f,t-\tau} (\overline{\ln TFP_{t-\tau}} - \ln TFP_{f,t-\tau})$

where S is the set of establishments that stayed in that industry from year t-τ to year t, N is the set of establishments that newly entered and X is the set of establishments that exited. TFP with an upper bar denotes the industry-average TFP level.

In a similar way, we define the industry-level labor productivity of a certain industry in year t as

$$\ln LP_t = \sum_{f\in F} \lambda_{f,t} \ln LP_{f,t} \tag{3.4}$$

where $\lambda_{f,t}$ denotes establishment f's labor input share in that industry in year t. Then we can decompose labor productivity growth in the manufacturing sector from year t-τ to year t, $lnLP_t - lnLP_{t-\tau}$, into the following five factors.

Within effect: $\sum_{f\in S} \lambda_{f,t-\tau} \Delta \ln LP_{f,t}$

Between effect: $\sum_{f\in S} \Delta\lambda_{f,t} (\ln LP_{f,t-\tau} - \overline{\ln LP_{t-\tau}})$

Covariance effect: $\sum_{f\in S} \Delta\lambda_{f,t} \Delta \ln LP_{f,t}$

Entry effect: $\sum_{f\in N} \lambda_{f,t} (\ln LP_{f,t} - \overline{\ln LP_{t-\tau}})$

Exit effect: $\sum_{f\in X} \lambda_{f,t-\tau} (\overline{\ln LP_{t-\tau}} - \ln LP_{f,t-\tau})$

where S is the set of establishments that stayed in that industry from year t-τ to year t, N is the set of establishments that newly entered and X is the set of establishments that exited. *LP* with an upper bar denotes the industry-average labor productivity level.

The results of our decomposition for the period from 1981 to 2003 (2000 in case of TFP decompositions) are reported in Table 3.3 and Figures 3.2 and 3.3. We divided the whole period into five sub-periods (four sub-periods in case of TFP decomposition): 1981-85, 85-90, 90-95, 95-2000, and 2000-2003. The switch-in and switch-out effects in Table 3.1 and Figures 3.2 and 3.3 show the contribution of those establishments that moved from one industry to another to the industry average of thc total factor and labor productivity level. It has been pointed out in preceding studies that decomposition results are affected by business cycles.[10] Figure 3.4 shows the capacity utilization rate and diffusion index of business conditions. In order to minimize the impact of business cycles on our estimation, we also decompose growth rates for the longer time spans of 1981-1990 and 1990-2003. The results are also reported in Table 3.2.

[10] In 1981-2002, there were four official business cycle peaks, June 1985, February 1991, May 1997, and November 2000, and five troughs, February 1983, November 1986, October 1993, January 1999, and January 2002. Official peak and trough dates are available in *Business Cycle Reference Dates*, Economic and Social Research Institute, Cabinet Office, Government of Japan (<http://www.esri.cao.go.jp/>).

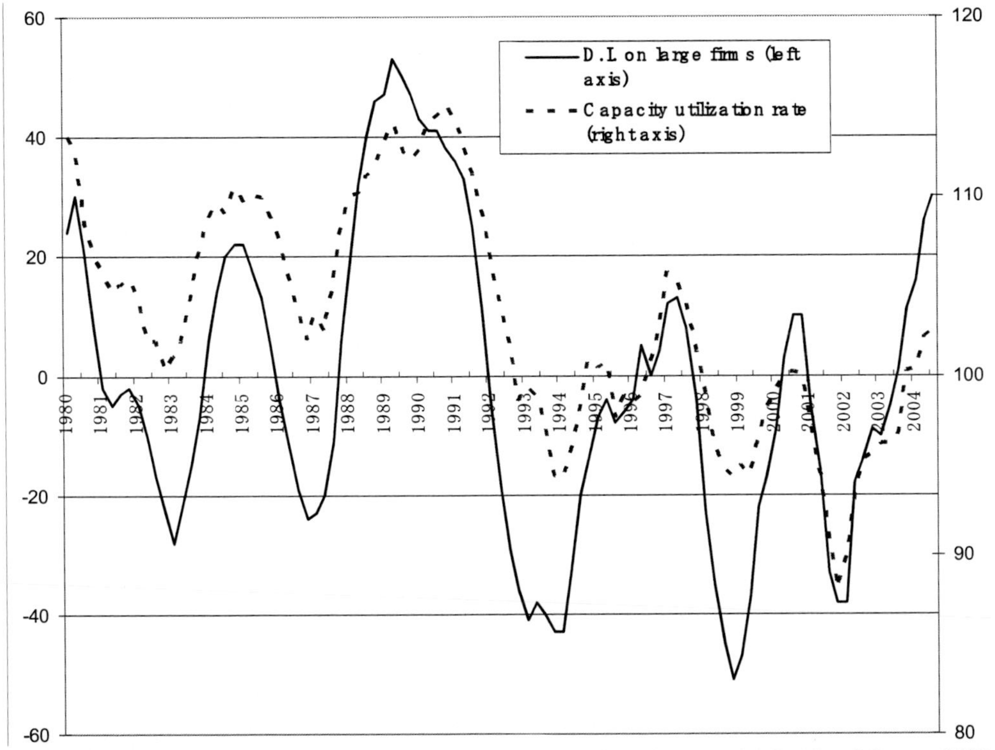

Sources: 'Tankan (Short-term Economic Survey of Enterprises in Japan),' Bank of Japan (diffusion index); METI (capacity utilization index).

Figure 3.4. Capacity Utilization Rate and Diffusion Index of Business Conditions ("Favorable" minus "Unfavorable," for Large Firms) in the Manufacturing Sector.

Our most important findings can be summarized as follows

1. The final column of Table 3.1 reports the exit effect (including the switch-out effect). The contribution of exit effect in whole periods was negative, irrespective of productivity indexes. What is more, the negative exit effect has been steadily growing. The negative exit effect means that the productivity level of exiting establishments has been higher than the industry average. We also found that the exit effect was negative for all industries. Fukao and Kwon (2006) also found similar negative exit effects using firm-level data in the 1990s.
2. The entry effect (including switch-in effect) was both positive and has tended to increase (Table 3.1). The entry effect was positive in almost all the industries. Contrary to exit of establishments, the entry of new establishments contributes to raising industry productivity growth.[11]
3. Moreover, the net entry effect has been positive in all periods. The entry and exit process tends to raise productivity growth as the entry of high productivity establishments exceeds the productivity of exit firms. The net contribution to overall

[11] Because of five and ten year intervals, we observe as entry establishments not start-up establishments but 2 years over establishments. Therefore, in the case of annual data, our entry effect is calculated as the productivity growth of surviving establishments.

productivity growth of the entry and exit of establishments is small and has declined over time. In the long-term, contribution of net entry effect to productivity growth is stable but gradually decreasing. It is noteworthy that the Japanese economy experienced the decline of the contribution of plant turnover to productivity growth in the 1990s.

4. The within effect, i.e. the effect of TFP growth within staying establishments, made the largest contribution to overall productivity growth in all periods (Table 3.1). However, this effect saw a sharp drop in the 1990s. The decline in the within effect is primarily attributable to the slowdown in productivity growth in the manufacturing sector.
5. The reallocation effect, which is the sum of the between effect and the covariance effect, contributed about half of the total productivity growth in the 1990s (in the case of TFP growth). This indicates that market forces have played an important role in efficiently allocating resources away from establishments that perform poorly and to establishments that perform well. In particular, the covariance effect, which gives the contribution of surviving establishments with increasing shares and growing productivity, seems to have been more important in the 1990s. This indicates that, instead of low net entry effect, the reallocation effect across existing establishments plays a stronger role in promoting productivity growth in the 1990s (Table 3.1). The reallocation effect is interpreted as reflecting the creative destruction processes (Bartelsman et al, 2004). The creative destruction within existing establishments in Japan in the 1990s was a very important factor in supporting aggregate productivity. In this context, it is encouraging that the Japanese government is trying to further strengthen the role of market forces through structural reforms.

In order to put Japan's productivity performance into an international perspective, we compare our results with those of studies for the U.S., Canada, the U.K., and South Korea. However, it should be born in mind that such a comparison presents various problems as the studies use different data sets, weights, and time spans. Below, we compare our results with those of studies for the countries mentioned that use the same methodology, although the time periods differ:

6. In recessionary periods, the reallocation effect rather than the within effect dominates. In contrast, during boom periods, it is the within effect that dominates. In Japan, however, the share of within effect does not change even in a great depression, although the growth of the within effect drastically dropped.
7. In all periods and all countries, the net entry effect is positive, suggesting that productivity grows as less productive establishments exit and more productive ones enter In addition, we find that in almost all countries except Japan, the net entry effect is the major source of overall productivity growth during recessionary periods. Yet, although Japan's recession lasted longer than those in other countries, we find that the contribution of the net entry effect is smallest, indicating that the mechanism of "creative destruction" in Japan's manufacturing sector is very weak.
8. The redistribution of market shares among incumbents, whereby high productivity establishments increase market shares and relatively low productivity decreases them, contributes positively to aggregate productivity growth in all periods and

countries. Overall, the productivity dynamics-sum of reallocation effect and net entry effect is larger than the within effect in all countries except Japan. This means that productivity growth due to market selection is essential to boost aggregate productivity growth. In Japan, productivity dynamics positively contributed to aggregate productivity growth, but the effect was smaller than that of other countries. This "low metabolism" seems to have slowed down the TFP growth of the manufacturing sector.

To sum up the above results, we find that the decline in Japan's TFP growth in the 1990s is largely due to the drop in the within effect. Conversely, the share of the reallocation effect has grown due to a strengthening of the role of the market. Yet, while the above analysis allowed us to determine the relative contribution of the within and the reallocation effect, it tells us little about why the within effect declined and why the negative exit effect increased.

4. Analysis of Causes of Negative Exit Effects and Sharp Decline in Within Effects

As shown in the previous section, the decline in Japan's TFP and labor productivity growth in the 1990s and the early 2000s was not caused by a slowdown of the "metabolism," i.e., the reallocation effect. Rather, it was primarily caused by a decline in the within effect. That is, in the period since 1990, it was the TFP and labor productivity growth within establishments that slowed down. We also showed that the absolute size of the negative exit effect became larger in the 1990s and the early 2000s. On a brighter note, we found that the positive entry effect increased during the period from 2000 to 2003. In order to understand how these changes in the 1990s and early 2000s came about, in this section we investigate the productivity dynamics and the entry and exit of establishments in greater detail.

Panel A of Figure 4.1 shows the share of establishments that were newly opened or closed in the total number of establishments.[12] Panel B shows the share accounted for by such establishments in total manufacturing output. These two figures show that the closure rate has been growing almost continuously since the beginning of the 1990s. On the other hand, the start-up rate declined in the 1990s, but then recovered during 2000-2003. The share of start-up establishments in total output did not decline in the 1990s because more larger establishments were opened in this period than in the 1980s.

[12] As we have already explained in the previous section, the Census covers all establishments in years ending with 0, 3, 5 and 8 of the Western calendar year. For other years, the Census covers establishments with four or more employees. In years when the Census covers only establishments with four or more employees, some establishments with more than four employees report that they employ less than four in order to evade the obligation to fill in the questionnaire. Because of such behavior, the number of establishments and our figures on "start-ups" tend to become higher in years when the Census covers all establishments. For more on this issue, see Shimizu and Miyagawa (2003).

Table 2.1. Macro-level Growth Accounting Results: 1970-2002 (annual rate of growth in %)

	Real GDP Growth	Contribution of Labor	Man-hour Growth	Labor Quality Growth	Contribution of Capital	Capital Stock Growth	Capital Quality Growth	Contribution of TFP Growth
	(a)	(b)=(c)+(d)	(c)	(d)	(e)=(f)+(g)	(f)	(g)	(h)=(a)-(b)-(e)
1970-75	5.47%	0.24%	-0.42%	0.66%	3.59%	2.94%	0.65%	1.64%
1975-80	5.69%	1.35%	0.87%	0.48%	1.98%	2.06%	-0.08%	2.37%
1980-85	3.92%	0.81%	0.31%	0.51%	2.12%	1.72%	0.40%	0.98%
1985-90	4.91%	0.68%	0.38%	0.30%	2.46%	1.87%	0.59%	1.77%
1990-95	1.45%	-0.01%	-0.41%	0.40%	1.41%	1.35%	0.05%	0.04%
1995-2000	1.27%	-0.06%	-0.42%	0.36%	0.92%	0.79%	0.13%	0.41%
2000-2002	-0.22%	-0.98%	-1.03%	0.04%	0.37%	0.31%	0.06%	0.39%
1970-80	5.58%	0.78%	0.24%	0.54%	3.17%	2.82%	0.36%	1.63%
1980-90	4.41%	0.75%	0.35%	0.41%	2.24%	1.76%	0.49%	1.42%
1990-2002	1.10%	-0.19%	-0.51%	0.33%	1.03%	0.94%	0.09%	0.25%

Source: JIP 2006 database.

Note: The growth accounting at the sector-level was conducted using the divisia index.

Table 2.2. Sector-level Growth Accounting Results: 1970-2002
(annual rate of growth in %)

	Real Output Growth	Intermediate Input Growth	Contribution of Labor	Man-hour Growth	Labor Quality Growth	Contribution of Capital	Capital Stock Growth	Capital Quality Growth	Contribution of TFP Growth
	(a)	(b)	(c)=(d)+(e)	(d)	(e)	(f)=(g)+(h)	(g)	(h)	(i)=(a)-(b)-(c)-(f)
Panel A. Manufacturing Sector									
1970-75	3.29%	1.75%	-0.28%	-0.44%	0.16%	0.57%	0.59%	-0.02%	1.25%
1975-80	5.11%	3.47%	0.32%	0.19%	0.13%	0.18%	0.16%	0.02%	1.13%
1980-85	4.08%	2.02%	0.36%	0.27%	0.09%	0.45%	0.33%	0.12%	1.25%
1985-90	4.59%	2.90%	0.05%	-0,05%	0.11%	0.62%	0.51%	0.11%	1.01%
1990-95	-0.09%	-0.30%	-0.46%	-0.60%	0.14%	0.39%	0.38%	0.01%	0.27%
1995-2000	0.57%	0.00%	-0.31%	-0.46%	0.16%	0.20%	0.14%	0.05%	0.68%
2000-2002	-2.73%	-194%	-0.73%	-0.83%	0.10%	-0.03%	-0.02%	0.00%	-0.03%
1970-80	4.20%	2.67%	0.02%	-0.11%	0.14%	0.36%	0.36%	0.00%	1.15%
1980-90	4.33%	2.48%	0.21%	0.11%	0.10%	0.51%	0.40%	0.11%	1.13%
1990-2002	-0.26%	-0.45%	-0.43%	-0.56%	0.13%	0.23%	0.21%	0.03%	0.39%
Panel B. Non-manufacturing Sector									
1970-75	4.72%	1.79%	0.42%	0.02%	0.40%	2.42%	1.85%	0.57%	0.09%
1975-80	4.94%	1.92%	0.83%	0.55%	0.28%	1.44%	1.47%	-0.03%	0.74%
1980-85	2.98%	1.34%	0.40%	0.06%	0.34%	1.36%	1.12%	0.23%	-0.11%
1985-90	4.69%	1.88%	0.49%	0.33%	0.16%	1.52%	1.15%	0.37%	0.80%
1990-95	2.24%	1.31%	0.25%	0.02%	0.22%	0.85%	0.82%	0.03%	-0.17%
1995-2000	1.13%	0.45%	0.11%	-0.09%	0.20%	0.58%	0.50%	0.08%	-0.02%
2000-2002	0.61%	0.38%	-0.35%	-0.35%	0.00%	0.27%	0.21%	0.06%	0.30%
1970-80	4.83%	1.89%	0.60%	0.28%	0.32%	2.25%	1.90%	0.35%	0.10%
1980-90	3.84%	1.63%	0.45%	0.20%	0.25%	1.39%	1.10%	0.29%	0.37%
1990-2002	1.50%	0.83%	0.09%	-0.09%	0.17%	0.63%	0.57%	0.06%	-0.04%

Source: JIP 2006 database.

Note: The growth accounting at the sector-level was conducted using the divisia index.

Table 3.1. Productivity Decomposition for the Manufacturing Sector as a Whole

	Annual productivity growth total (%)	Within share (%)	Reallocation share subtotal (%)	Between share (%)	Covariance share (%)	Net entry share (%)	Entry share (%)	Exit share (%)
	a=b+c	B	c=d+e+f	d	e	f=g+h	g	h
	A. TFP Growth							
1981-90	1.81	1.18	0.62	-0.14	0.28	0.49	0.73	-0.24
		(65.5)	(34.5)	(-8.0)	(15.3)	(27.2)	(40.2)	(-13.1)
1990-2003	1.12	0.55	0.58	-0.04	0.35	0.27	0.60	-0.33
		(48.8)	(51.2)	(-3.4)	(30.7)	(23.9)	(53.1)	(-29.3)
1981-85	1.49	1.12	0.37	-0.29	0.50	0.16	0.49	-0.33
		(75.0)	(25.0)	(-19.1)	(33.6)	(10.6)	(32.9)	(-22.3)
1985-90	2.07	1.46	0.61	-0.24	0.55	0.30	0.59	-0.29
		(70.5)	(29.5)	(-11.4)	(26.5)	(14.3)	(28.3)	(-14.0)
1990-95	1.51	1.02	0.50	-0.23	0.56	0.17	0.44	-0.26
		(67.1)	(32.9)	(-15.4)	(36.8)	(11.5)	(28.8)	(-17.3)
1995-2000	1.09	0.57	0.52	-0.14	0.59	0.07	0.47	-0.40
		(52.2)	(47.8)	(-13.0)	(54.5)	(6.3)	(42.7)	(-36.4)
2000-2003	0.90	0.34	0.57	-0.15	0.78	-0.07	0.79	-0.86
		(37.2)	(62.8)	(-16.2)	(86.6)	(-7.6)	(87.6)	(-95.2)
	B. Labor Productivity Growth							
1981-90	4.44	3,.34	1.10	-0.01	-0.45	1.56	1.97	-0.41
		(75.2)	(24.8)	(-0.2)	(-10.2)	(35.2)	(44.4)	(-9.2)
1990-2003	2.41	1.15	1.26	0.30	-0.02	0.98	1.54	-0.56
		(47.7)	(52.3)	(12.4)	(-0.9)	(40.9)	(64.1)	(-23.3)
1981-85	3.65	3.60	0.04	0.01	-0.64	0.67	1.23	-0.56
		(98.8)	(1.2)	(0.3)	(-17.5)	(18.4)	(33.7)	(-15.3)
1985-90	5.39	4.50	0.89	0.19	-0.60	1.29	1.69	-0.40
		(83.5)	(16.5)	(3.6)	(-11.1)	(24.0)	(31.3)	(-7.3)
1990-95	3.75	2.79	0.96	0.51	-0.41	0.86	1.23	-0.37
		(74.4)	(25.6)	(13.6)	(-11.0)	(23.0)	(32.9)	(-9.9)
1995-2000	1.69	1.26	0.44	0.38	-0.34	0.39	0.94	-0.55
		(74.2)	(25.8)	(22.6)	(-20.1)	(23.3)	(55.7)	(-32.5)
2000-2003	1.61	1.31	0.31	1.12	-0.95	0.13	1.56	-1.43
		(81.0)	(19.0)	(69.7)	(-58.6)	(8.0)	(96.4)	(-88.4)

Notes: Values in parentheses denote the share of each effect in aggregate productivity growth.

Table 3.2. Comparison of the Decomposition of Total Factor Productivity for Various Countries

Source	Country	Period	Annual productivity growth total (%)	Contribution of each effect						
				Within effect(%)	Reallocation effect subtotal(%)	Between effect(%)	Covariance effect(%)	Net entry effect subtotal(%)	Entry effect (%)	Exit effect(%)
			a=b+c+f	b	c=d+e	d	e	f=g+h	g	h
Panel A. TFP Growth										
Ahn, Kwon, Fukao (2005)	South Korea	1990-98	3.51	1.42	0.08	-0.28	0.36	2.01	1.95	0.06
				(0.40)	(0.02)	(-0.08)	(0.10)	(0.57)	(0.56)	(0.02)
Foster, Haltiwanger, and Krizan (2001)	U.S	1977-87	1.02	0.49	0.27	-0.08	0.35	0.27		
				(0.48)	(0.26)	(-0.08)	(0.34)	(0.26)		
		1977-82	0.54	-0.05	0.45	-0.18	0.63	0.14		
				(-0.09)	(0.83)	(-0.33)	(1.16)	(0.25)		
		1982-87	1.46	0.76	0.48	-0.26	0.75	0.20		
				(0.52)	(0.33)	(-0.18)	(0.51)	(0.14)		
		1987-92	0.66	-0.04	0.47	-0.26	0.73	0.23		
				(-0.06)	(0.71)	(-0.39)	(1.10)	(0.35)		
Disney , Haskel, and Heden (2003)	U.K.	1980-92	1.06	0.05	0.43	0.16	0.28	0.57		
				(0.05)	(0.41)	(0,15)	(0.26)	(0.54)		
		1982-87	3.08	1.26	1.48	-0.09	1.57	0.37		
				(0.41)	(0.48)	(-0.03)	(0.51)	(0.12)		
This paper	Japan	1981-90	1.81	1.18	0.13	-0.14	0.28	0.49	0.73	-0.24
				(0.66)	(0.07)	(-0.08)	(0.15)	(0.27)	(0.40)	(-0.13)
		1990-2000	1.27	0.72	0.29	-0.10	0.40	0.26	0.54	-0.29
				(0.57)	(0.23)	(-0.08)	(0.31)	(0.20)	(0.43)	(-0.23)

Table 3.2. (Continued)

Source	Country	Period	Annual productivity growth total (%)	Contribution of each effect						
				Within effect(%)	Reallocation effect subtotal(%)	Between effect(%)	Covariance effect(%)	Net entry effect subtotal(%)	Entry effect (%)	Exit effect(%)
			a=b+c+f	b	c=d+e	d	e	f=g+h	g	h
Panel B. Labor Productivity Growth										
Foster, Haltiwanger, and Krizan (2001)	U.S.	1977-87	2.13	1.64	-0.13	0.17	-0.30	0.62		
				(0.77)	(0.06)	(0.08)	(-0.14)	(0.29)		
		1977-82	0.51	0.62	-0.22	0.43	-0.65	0.10		
				(1.22)	(-0.22)	(0.85)	(-1.27)	(0.20)		
		1982-87	3.73	3.10	-0.07	0.49	-0.56	0.71		
				(0.83)	(0.02)	(0.13)	(-0.15)	(0.19)		
		1987-92	1.43	1.34	-0.23	0.47	-0.70	0.30		
				(0.94)	(0.16)	(0.33)	(-0.49)	(0.21)		
Baldwin and Gu (2003)	Canada	1973-79	2.15	1.66	-0.05	1.47	-1.52	0.54	0.24	0.30
				(0.77)	(0.03)	(0.68)	(-0.71)	(0.25)	(0.11)	(0.14)
		1979-88	1.41	1.44	-0.30	0.23	-0.53	0.28	0.15	0.13
				(1.02)	(0.22)	(0.16)	(-0.38)	(0.20)	(0.11)	(0.09)
		1988-97	2.91	2.85	-0.37	0.27	-0.64	0.42	0.26	0.17
				(0.98)	(0.13)	(0.09)	(-0.22)	(0.15)	(0.09)	(0,06)
This paper	Japan	1981-90	4.44	3.34	-0.46	-0.01	-0.45	1.56	1.97	-0.41
				(0.75)	(-0.10)	(-0.002)	(-0.10)	(0.35)	(0.44)	(-0.09)
		1990-2003	2.41	1.15	0.28	0.30	-0.02	0.98	1.54	-0.56
				(0.48)	(0.11)	(0.12)	(-0.01)	(0.41)	(0.64)	(-0.23)

Notes: The entry and exit effects in this paper and in Ahn, Kwon, and Fukao (2004) include the switch-in and switch-out effects. Values in parentheses denote the share of each effect in aggregate productivity growth. The decomposition of TFP for each country is based on the method of Foster, Haltiwanger and Krizan.

Table 4.1. Transition Matrices of Establishments' Rank in Labor Productivity: 1981-1990 and 1990-2003

1981-1990

		1990										Closed establish-ments	Total	
		Top LP group	2nd LP level group	3rd LP level group	4th LP level group	5th LP level group	6th LP level group	7th LP level group	8th LP level group	9th LP level group	Lowest LP group			
1981	Top LP group	11 741	5 886	3 376	2 259	1 636	1 230	963	748	619	570	13 086	42 114	421 367
	2nd LP level group	5 572	6 686	5 119	3 630	2 653	1 823	1 472	1 075	782	615	12 712	42 139	
	3rd LP level group	3 266	5 039	5 320	4 501	3 479	2 626	1 909	1 458	983	755	12 795	42 131	
	4th LP level group	2 174	3 533	4 510	4 708	4 207	3 400	2 554	1 802	1 288	887	13 079	42 142	
	5th LP level group	1 527	2 555	3 458	4 139	4 392	3 975	3 212	2 481	1 708	1 094	13 599	42 140	
	6th LP level group	1 218	1 808	2 524	3 362	3 894	4 293	3 910	3 264	2 350	1 433	14 080	42 136	
	7th LP level group	925	1 332	1 888	2 536	3 202	3 888	4 251	4 000	3 256	1 955	14 898	42 131	
	8th LP level group	706	976	1 331	1 840	2 521	3 258	3 963	4 522	4 279	2 863	15 881	42 140	
	9th LP level group	541	737	970	1 304	1 688	2 379	3 158	4 391	5 547	4 536	16 882	42 133	
	Lowest LP group	544	615	793	883	1 167	1 456	2 057	2 990	4 549	7 639	19 468	42 161	
Newly set-up establishments		13 956	13 028	12 897	13 030	13 362	13 856	14 738	15 464	16 822	19 872		147 025	
Total		42 170	42 195	42 186	42 192	42 201	42 184	42 187	42 195	42 183	42 219	146 480		
											421 912			

Table 4.1. (Continued)

1990-2003

		2003										Closed establish-ments	Total	
		Top LP group	2nd LP level group	3rd LP level group	4th LP level group	5th LP level group	6th LP level group	7th LP level group	8th LP level group	9th LP level group	Lowest LP group			
1990	Top LP group	8 137	4 472	2 887	1 915	1 399	1 064	852	648	544	511	20 007	42 436	424 535
	2nd LP level group	3 583	4 508	3 877	3 044	2 317	1 671	1 337	984	691	582	19 854	42 448	
	3rd LP level group	2 028	3 325	3 571	3 259	2 804	2 256	1 678	1 226	924	708	20 678	42 457	
	4th LP level group	1 323	2 250	2 808	3 047	2 915	2 515	2 107	1 659	1 256	839	21 735	42 454	
	5th LP level group	952	1 603	2 097	2 598	2 801	2 683	2 505	1 949	1 491	1 024	22 757	42 460	
	6th LP level group	737	1 062	1 612	1 986	2 332	2 732	2 684	2 387	1 881	1 201	23 834	42 448	
	7th LP level group	534	786	1 097	1 534	1 954	2 348	2 629	2 636	2 279	1 590	25 063	42 450	
	8th LP level group	400	608	787	1 040	1 393	1 913	2 367	2 718	2 793	2 080	26 360	42 459	
	9th LP level group	333	399	576	729	949	1 242	1 701	2 484	3 034	2 824	28 177	42 448	
	Lowest LP group	319	348	409	518	588	799	1 028	1 469	2 354	3 626	31 017	42 475	
Newly set-up establishments		10 255	9 263	8 897	8 953	9 174	9 392	9 728	10 466	11 370	13 654		101 152	
Total		28 601	28 624	28 618	28 623	28 626	28 615	28 616	28 626	28 617	28 639	239 482		
											286 205			

Table 4.2. Survival Rate after Entry

Entry Year	1981	1982	1983	1984	1985	1986	1987	1988	1989	1990	1991	1992	1993	1994	1995	1996	1997	1998	1999	2000	2001	2002	2003
1981 and before	100.0%	90.7%	86.0%	81.2%	78.4%	74.2%	70.7%	69.5%	66.1%	65.2%	62.1%	59.5%	57.8%	53.7%	52.8%	49.9%	47.6%	45.6%	42.4%	40.6%	37.4%	34.4%	31.7%
1982		100.0%	79.4%	71.6%	68.5%	62.2%	58.4%	57.4%	53.9%	53.1%	50.2%	47.5%	45.9%	41.9%	41.2%	38.8%	36.6%	34.8%	32,0%	30.4%	27.7%	25.2%	23.0%
1983			100.0%	74.6%	69.9%	59.5%	53.8%	55.4%	49.4%	50.7%	45.7%	42.4%	42.0%	36.8%	37.3%	33.7%	31.5%	30.8%	27.3%	26.3%	23.3%	20.8%	18.8%
1984				100.0%	78.9%	68.7%	63.7%	61.8%	57.7%	57.2%	54.2%	50.5%	48.2%	44.2%	43.2%	40.3%	38.4%	36.4%	33.8%	32.1%	29.2%	26.8%	24.4%
1985					100.0%	71.2%	62.5%	62.7%	56.3%	57.4%	52.1%	48.4%	47.5%	41.9%	42.3%	38.3%	36.1%	35.3%	32.0%	30.6%	27.1%	24.7%	22.5%
1986						100.0%	84.1%	78.0%	72.5%	70.1%	65.7%	61.8%	58.8%	53.5%	52.1%	48.5%	45.9%	43.4%	40.0%	37.9%	34.4%	31.2%	28.4%
1987							100.0%	80.2%	73.7%	70.6%	66.0%	61.5%	57.9%	52.8%	51.5%	48.2%	45.4%	43.0%	39.1%	37.1%	33.7%	30.6%	28.0%
1988								100.0%	70.5%	71.7%	62.7%	56.8%	56.2%	48.3%	49.1%	44.0%	41.0%	39.9%	35.2%	33.9%	29.8%	26.8%	24.1%
1989									100.0%	84.4%	77.5%	71.8%	66.6%	59.2%	57.8%	53.6%	50.4%	47.2%	43.7%	41.2%	37.2%	34.0%	30.8%
1990										100.0%	73.8%	65.5%	63.2%	53.9%	54.4%	49.0%	46.1%	45.4%	40.1%	38.6%	34.0%	30.8%	27.8%
1991											100.0%	82.4%	74.8%	66.2%	64.2%	59.6%	55.7%	52.2%	47.6%	44.9%	40.5%	36.6%	33.3%
1992												100.0%	79.2%	69.8%	67.9%	62.4%	58.2%	54.0%	49.5%	46.8%	42.4%	38.4%	35.4%
1993													100.0%	69.3%	67.9%	59.2%	54.1%	52.6%	46.2%	44.4%	38.7%	34.5%	31.1%
1994														100.0%	79.7%	71.6%	66.2%	61.9%	56.4%	52.7%	47.4%	43.2%	39.1%
1995															100.0%	74.1%	67.2%	63.2%	55.0%	52.2%	45.7%	40.7%	36.5%
1996																100.0%	81.2%	72.9%	65.5%	61.6%	55.5%	49.4%	44.0%
1997																	100.0%	79.5%	70.8%	66.5%	59.0%	52.3%	47.1%
1998																		100.0%	77.0%	71.8%	62.,6%	54.5%	48.1%
1999																			100.0%	77.8%	68.,2%	58.7%	51.0%
2000																				100.0%	64.1%	54.3%	44.7%
2001																					100.0%	70.2%	58.8%
2002																						100.0%	79.3%
2003																							100.0%

Table 4.3. Ratio of the Average Plant Size of Establishments Set Up in the Same Year to the Industry Average Plant Size: By Start-up Year and by Year of Observation

Entry Year	1981	1982	1983	1984	1985	1986	1987	1988	1989	1990	1991	1992	1993	1994	1995	1996	1997	1998	1999	2000	2001	2002	2003
1981 and before	100.0%	111.9%	120.5%	121.4%	128.6%	133.4%	133.6%	138.9%	139.3%	143.2%	146.7%	146.4%	147.6%	146.6%	149.0%	149.1%	150.4%	158.8%	156.2%	158.3%	156.4%	154.8%	159.6%
1982		41.9%	52.3%	56.6%	61.8%	65.2%	66.2%	70.1%	69.6%	71.7%	72.9%	73.8%	75.1%	73.6%	74.2%	74.5%	76.1%	81.8%	81.1%	81.0%	80.5%	83.8%	88.7%
1983			34.0%	42.8%	46.3%	52.9%	54.0%	53.7%	57.2%	57.6%	58.7%	61.3%	62.3%	63.3%	63.0%	66.1%	66.4%	70.1%	72.4%	70.7%	73.7%	75.4%	81.2%
1984				43.9%	60.0%	65.1%	66.1%	77.1%	79.2%	83.8%	84.0%	85.7%	85.1%	85.9%	88.5%	88.2%	88.5%	97.4%	101.2%	99.0%	101.1%	91.4%	98.9%
1985					40.4%	55.9%	61.8%	64.8%	69.3%	73.7%	79.4%	82.8%	85.5%	88.5%	86.5%	89.2%	89.8%	93.6%	97.1%	101.3%	101.3%	97.3%	98.6%
1986						40.5%	46.2%	51.4%	51.9%	56.2%	58.7%	58.5%	59.2%	57.8%	59.3%	60.2%	61.0%	65.4%	64.7%	64.6%	62.8%	62.9%	67.1%
1987							43.4%	57.8%	63.1%	68.9%	70.8%	69.8%	75.1%	75.4%	79.3%	79.9%	80.4%	87.4%	87.3%	89.4%	89.1%	88.5%	95.3%
1988								33.3%	43.6%	46.8%	52.1%	52.2%	54.2%	55.6%	56.2%	59.0%	61.9%	66.3%	68.3%	71.9%	75.4%	78.7%	82.4%
1989									39.9%	52.8%	58.9%	60.8%	68.2%	69.4%	73.4%	77.4%	79.9%	87.8%	84.0%	90.4%	87.1%	95.1%	86.6%
1990										36.0%	49.2%	53.1%	58.0%	61.9%	62.1%	67.4%	66.5%	68.6%	70.7%	72.4%	77.0%	81.3%	88.7%
1991											37.6%	46.4%	52.1%	55.1%	55.4%	59.4%	59.8%	64.6%	65.4%	66.2%	67.2%	68.0%	74.4%
1992												47.1%	52.2%	59.6%	65.4%	61.9%	68.7%	73.2%	80.5%	80.5%	79.2%	84.7%	96.3%
1993													33.8%	45.9%	51.3%	53.8%	55.4%	58.6%	63.0%	67.9%	73.4%	73.8%	78.9%
1994														47.5%	56.5%	60.7%	63.5%	71.8%	71.5%	75.7%	77.8%	85.2%	90.4%
1995															42.1%	56.7%	60.5%	68.9%	71.3%	74.8%	72.5%	70.4%	75.7%
1996																41.1%	49.4%	57.9%	62.6%	70.2%	69.3%	70.0%	79.4%
1997																	40.2%	55.6%	57.7%	61.5%	65.0%	67.2%	79.2%
1998																		35.0%	40.7%	41.6%	42.4%	42.8%	49.1%
1999																			50.1%	61.5%	63.6%	66.9%	74.4%
2000																				36.7%	49.7%	55.7%	63.5%
2001																					53.0%	69.1%	82.1%
2002																						63.1%	84.6%
2003																							41.3%

Note: Establishment size is measured by nominal output.

Table 4.4. Ratio of the Average Plant Size of Establishments Closed in the Same Year to the Industry Average Plant Size: By Closure Year and by Year of Observation

Entry Year	1981	1982	1983	1984	1985	1986	1987	1988	1989	1990	1991	1992	1993	1994	1995	1996	1997	1998	1999	2000	2001	2002	2003
	100.0%	41.9%	34.0%	43.9%	40.4%	40.5%	43.4%	33.3%	39.9%	36.0%	37.6%	47.1%	33.8%	47.5%	42.1%	41.1%	40.2%	35.0%	50.1%	36.7%	53.0%	63.1%	41.3%
1981	36.4%																						
1982	46.8%	40.5%																					
1983	67.7%	55.7%	43.2%																				
1984	53.3%	50.3%	47.1%	40.2%																			
1985	59.5%	56.3%	50.1%	48.4%	38.4%																		
1986	58.5%	56.2%	50.9%	49.6%	42.2%	36.8%																	
1987	60.1%	59.0%	55.7%	49.3%	50.0%	44.6%	39.0%																
1988	64.4%	61.6%	58.5%	57.8%	53.3%	45.5%	42.0%	37.0%															
1989	73.3%	69.0%	65.2%	62.5%	56.9%	52.2%	49.8%	46.2%	38.0%														
1990	55.9%	54.7%	53.8%	52.3%	49.3%	46.8%	44.8%	41.5%	39.7%	32.4%													
1991	49.1%	47.2%	46.8%	47.8%	45.6%	44.4%	43.3%	41.8%	40.3%	37.7%	33.0%												
1992	62.4%	60.9%	57.5%	54.9%	52.6%	51.5%	48.9%	47.8%	43.7%	41.1%	39.4%	34.2%											
1993	68.4%	66.3%	65.0%	62.2%	59.2%	58.5%	56.0%	52.2%	51.2%	49.2%	47.0%	45.3%	37.0%										
1994	75.2%	75.3%	69.8%	69.5%	65.6%	64.1%	59.9%	58.3%	56.7%	53.7%	49.4%	49.0%	42.7%	40.0%									
1995	76.8%	72.5%	67.8%	65.3%	62.7%	59.4%	56.6%	53.6%	53.4%	51.4%	50.4%	52.0%	44.5%	41.5%	34.2%								
1996	80.0%	78.2%	76.2%	71.2%	68.8%	65.5%	62.5%	58.7%	54.9%	57.8%	56.1%	53.0%	53.3%	45.3%	42.2%	36.4%							
1997	83.4%	80.6%	79.3%	73.8%	73.3%	68.9%	66.5%	63.7%	60.4%	58.4%	56.3%	56.8%	56.0%	51.2%	47.2%	43.8%	35.9%						
1998	89.7%	85.1%	83.4%	82.0%	80.8%	77.4%	74.2%	72.8%	70.6%	69.0%	67.2%	63.1%	59.6%	58.0%	54.0%	52.0%	49.9%	41.1%					
1999	124.5%	142.9%	123.1%	117.2%	113.5%	107.7%	101.0%	90.5%	84.4%	79.3%	82.3%	78.6%	67.0%	66.5%	63.4%	58.7%	55.3%	49.1%	42.4%				
2000	91.1%	89.3%	86.5%	82.5%	82.2%	78.3%	75.3%	73.6%	71.0%	69.7%	68.7%	65.8%	62.4%	59.8%	58.3%	58.9%	56.1%	52.3%	49.1%	41.2%			
2001	114.1%	107.3%	108.4%	107.1%	101.5%	96.4%	95.4%	95.2%	92.9%	93.4%	89.0%	84.6%	81.9%	76.9%	74.5%	70.4%	67.6%	63.8%	60.4%	55.9%	47.0%		
2002	143.7%	140.2%	135.1%	134.1%	137.2%	124.5%	118.0%	117.7%	113.8%	110.9%	106.4%	102.5%	101.2%	93.3%	89.9%	84.1%	80.3%	75.4%	68.8%	65.8%	60.3%	51.5%	
2003 and after	199.6%	193.2%	194.0%	188.0%	186.0%	178.9%	172.1%	173.1%	167.8%	167.9%	163.3%	157.7%	155.6%	148.3%	146.8%	140.7%	136.8%	135.3%	128.1%	124.7%	117.4%	108.1%	100.0%

Note: Establishment size is measured by nominal output.

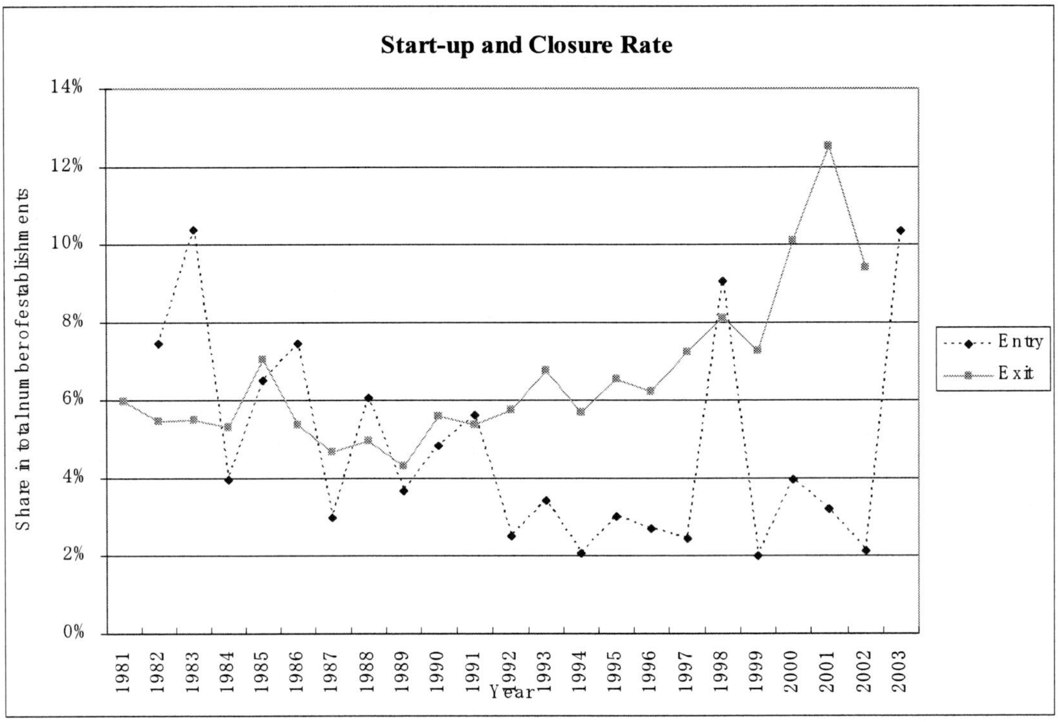

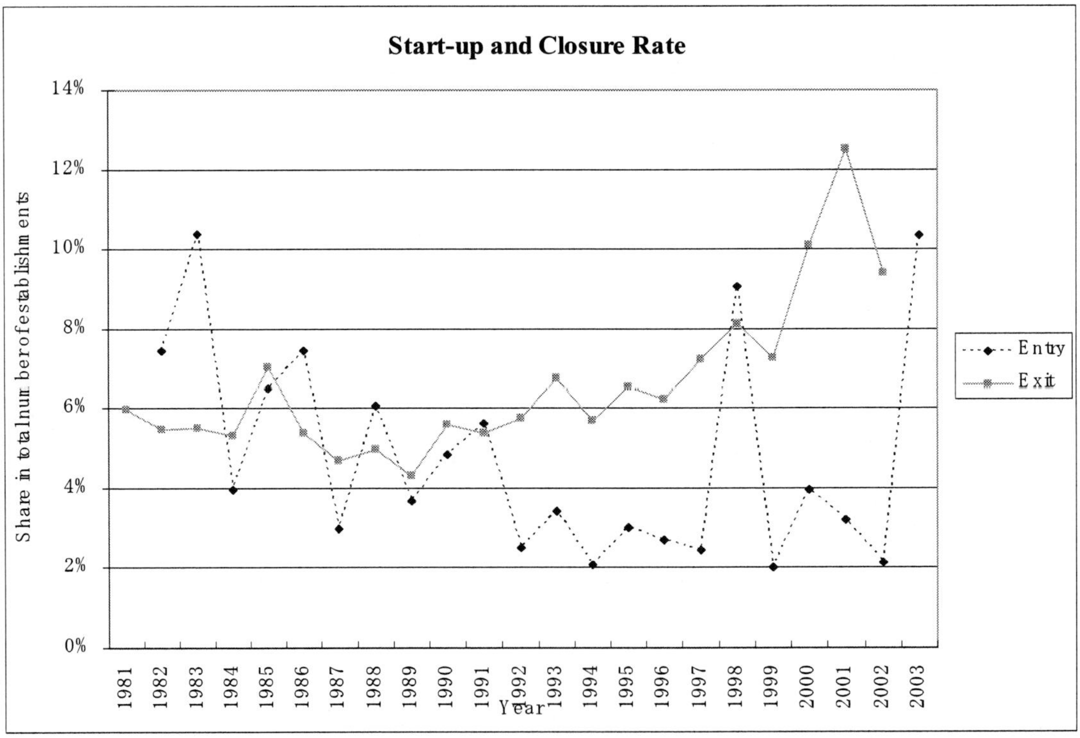

Figure 4.1. Share of Newly Opened and Closed Establishments in Total Establishments.

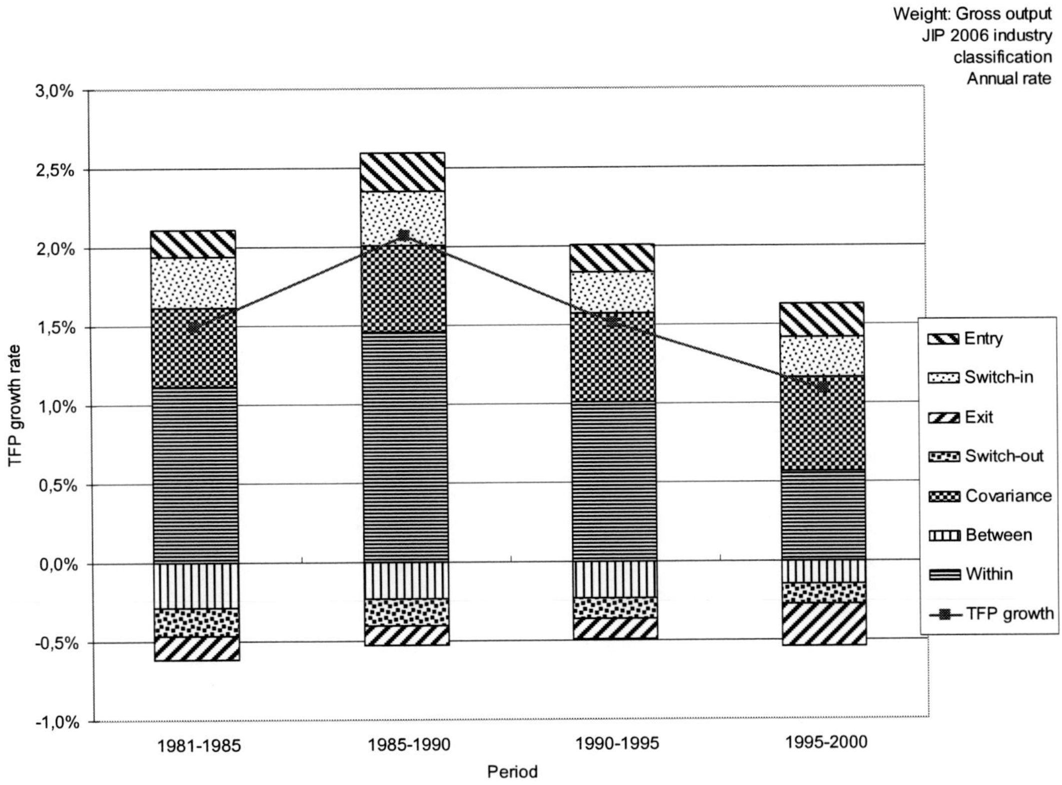

Figure 3.2. Decomposition of TFP Growth.

Many establishments could not survive in the period 1990-2003. We can confirm this by looking at the transition matrices of establishments' rank in labor productivity (Table 4.1). Only 44% of all the establishments which existed in 1990 survived until 2003 (the average annual survival rate was 93.8%). In contrast, from 1981 to 1990, 65% establishments continued to operate (the annual survival rate was 95.4%). In the 1990-2003 period, 239,482 establishments were closed and only 101,152 establishments were opened. As a result, the number of establishments declined by 33% from 424,535 to 286,205. In the 1980-1990 period, the number of closures and the number of start-ups were more or less in balance and the number of establishments actually increased slightly.

Table 4.1 also shows that the degree of persistence of labor productivity is very high. 56% of establishments which originally ranked in the bottom three deciles in 1990 remained in the same three deciles in 2003. Similarly, 55% of establishments which originally ranked in the top three deciles in 1990 remained in the same three deciles in 2003 (if they survived). We also found a similar persistence of productivity in the case of TFP.[16]

[16] Several studies examining plant level TFP found the degree of persistence to be also very high in the U.S. manufacturing sector (Baily *et al.*, 1992; Bartelsman and Doms, 2000). Fukao and Kwon (2006) found a similar high persistence in firm-level TFP in Japan's manufacturing sector.

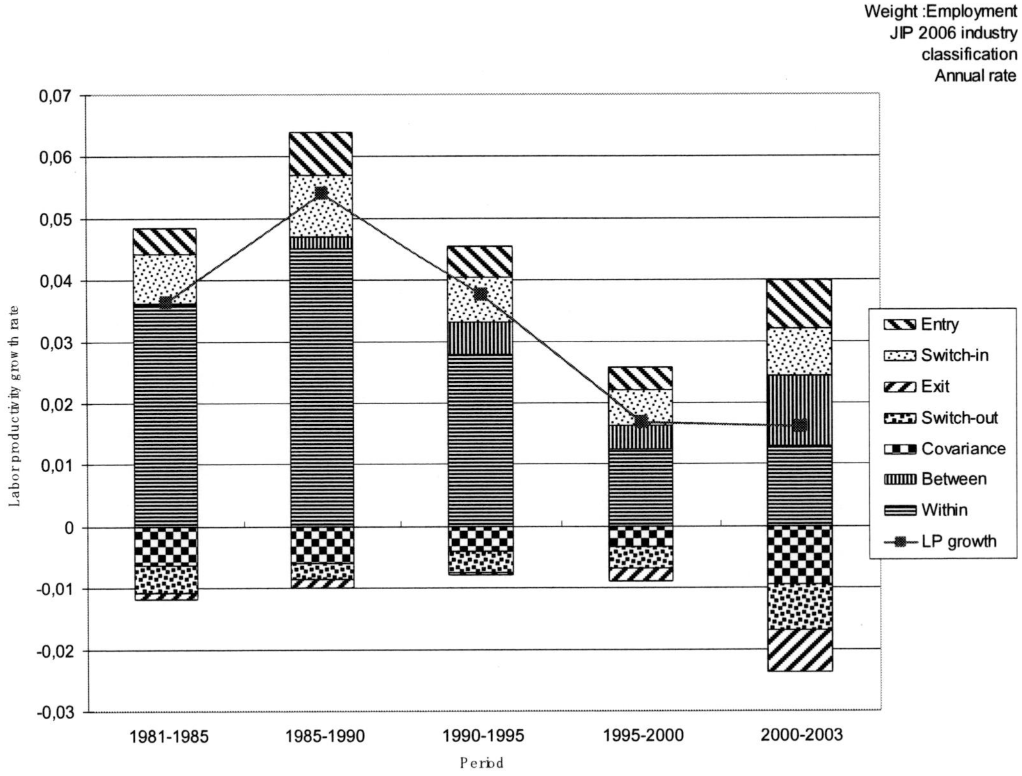

Figure 3.3. Decomposition of Labor Productivity Growth.

It is important to note that the survival rate is not high even in the case of establishments in the top labor-productivity group. From 1990 to 2003, only 47% of establishments in this group survived. The survival rates of establishments in the higher labor productivity groups are slightly greater than those of establishments in the lower labor productivity groups. But the average size of establishments in the higher labor productivity groups is larger than in the lower labor productivity groups. These factors cause the negative exit effect that can be seen in Figure 3.2 and Figure 3.3.[17] As these figures show, the negative exit effect worsened after 1995.

Why did the negative exit effect worsen after 1995? One possible explanation is a hollowing-out effect as a result of direct investment abroad. In the 1990s, Japanese firms relocated production to Asian countries (primarily the ASEAN countries and China) in order to lower production costs. As Figure 4.2 shows, in the case of the electrical machinery industry, there was indeed a very rapid increase in production abroad and a decline in domestic production and net exports in the period 1990-2003. Since it is mainly productive large firms that invest abroad, this relocation of production may have led to the closure of productive establishments in Japan.

[17] Using firm-level TFP data for Japan's manufacturing sector for the period after 1994, Nishimura, Nakajima and Kiyota (2005) and Fukao and Kwon (2006) found similar negative exit effects.

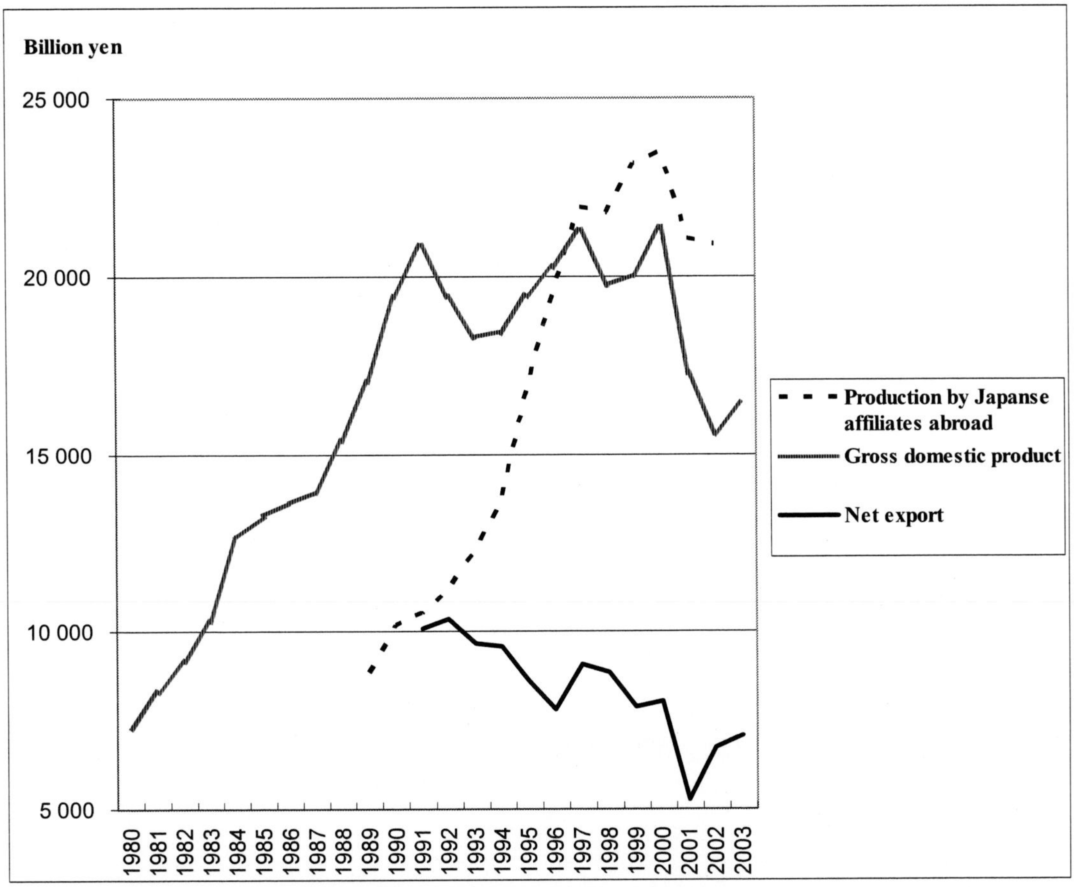

Sources: Cabinet Office, *Annual Report on National Accounts, 2006* RIETI's Database on Japan's Direct Investment Abroad, downloaded from http://www.rieti.go.jp/jp/database/d02.html#01 on July 15, 2006.

Figure 4.2. Gross Domestic Product and Production Abroad by Japan's Electrical Machinery Industry.

If we can link our establishment data with firm-level data of direct investment abroad, we could test the above hypothesis. Unfortunately, we do not have such linked micro data at this moment. But at the industry level, we can test our hypothesis. Figure 4.3 compares changes in Japanese firms' production in East Asia (China, Korea, Taiwan, Hong Kong, Indonesia, Thailand, the Philippines, and Malaysia) from 1990 to 2002 with the exit effect in 1990-2003. We have 48 cross industry observations. Consistent with our hypothesis, we can observe a negative correlation between production in Asia and the exit effect. The correlation coefficient is 0.42, which is statistically significant at the 5% level. When we exclude observations for the communication equipment industry, which seems to contribute strongly to the negative correlation, the correlation coefficient declines to 0.24 but is still significant at the 5% level.

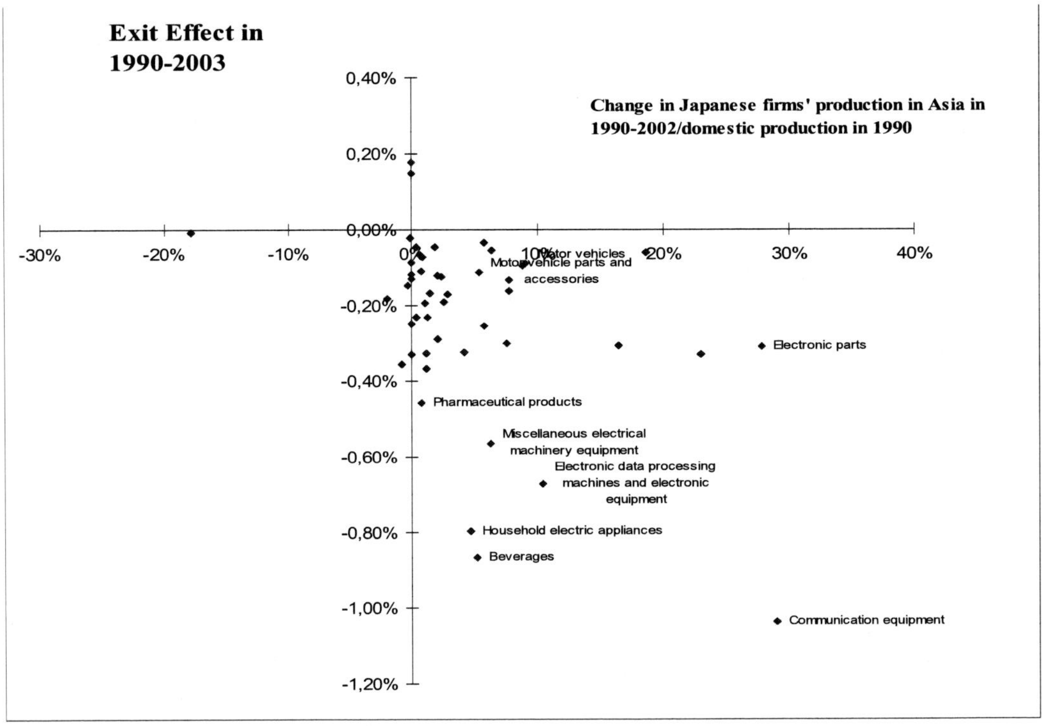

Source: Data on Japanese firms' production in East Asia is from the JIP 2006.

Figure 4.3. Direct Investment in East Asia and the Exit Effect in 1990-2003.

Based on a different approach but using similar longitudinal labor productivity data on manufacturing establishments (for the period 1985-95) , Shimizu and Miyagawa (2003) found that in many industries, such as the mechanical rubber products and household electric appliances industries, the labor productivity of establishments that closed was significantly higher than that of establishments that stayed. Although they focused on an earlier period than ours, they arrived at similar inferences as we did, that is, the closure of productive establishments may be caused by the relocation of production from Japan to foreign countries.

Although we need linked micro data in order to draw conclusions about the cause of the worsening negative exit effect, the hollowing-out hypothesis seems to be a leading hypothesis.

Next, let us analyze about the slowdown of within effects in Japan's manufacturing from the 1990s. In order to completely understand the trend of within effect, we need firm level data such as, R&D, IT investment at head office, M&A, international trade and out-sourcing etc. At this moment we do not have such data. But using the data we have, we can analyze how a sharp decline in the number of establishments contributed to the slow-down of within effects.

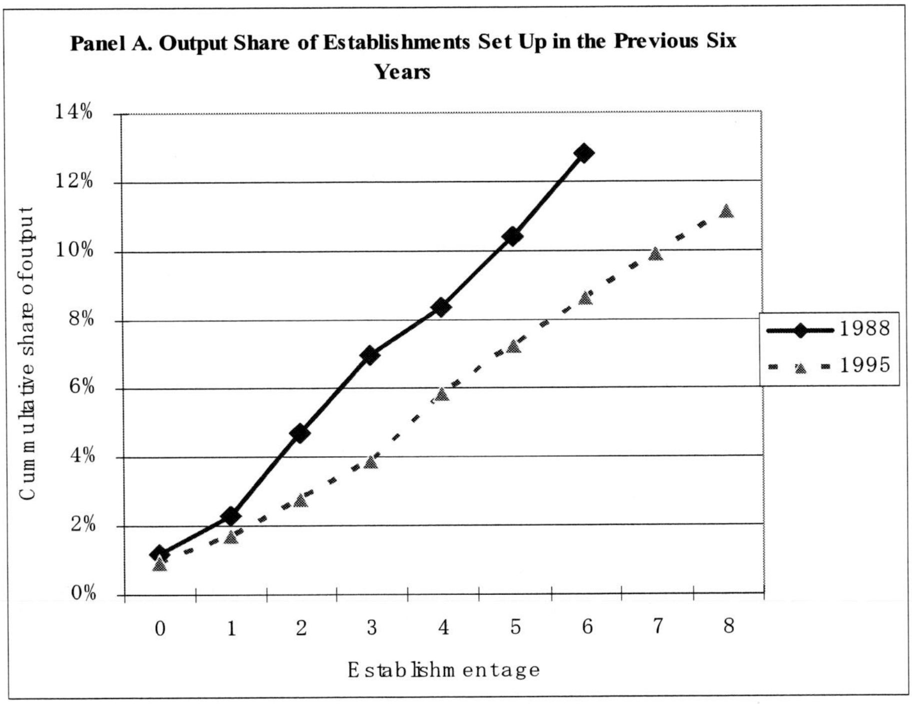

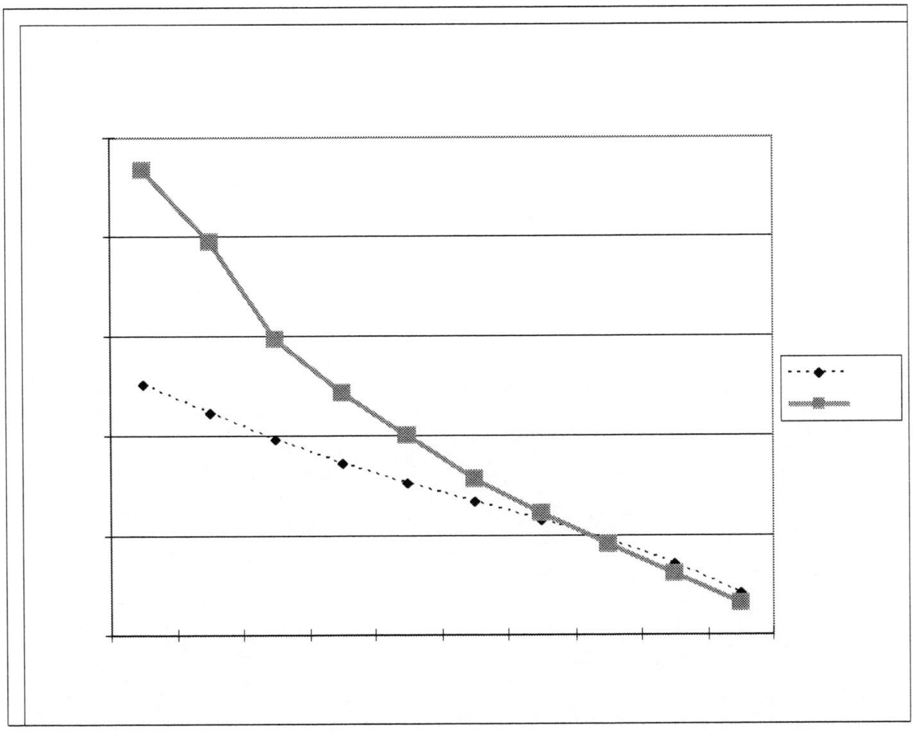

Figure 4.4. Output Share of New Establishments and Establishments Closing in the Near Future.

According to proceeding studies, new establishments gradually expand their size and improve their productivity over time (Baldwin 1998, Bartelsman and Doms 2000) and exiting firms have lower performance several years before their exit (Kiyota and Takizawa 2006). In our decomposition analysis some part of such learning effects and shadow-of-death effects might be included in our within effect. For example, suppose one establishment is opened in 1983 and gradually improves its productivity from 1983 to 1990. If the productivity level of this establishment has already become high by 1985, this fact is counted as a positive entry effect for the period of 1980-1985. But its productivity improvement after 1985 is counted as within effect for the period of 1985-1990. In a similar way some part of shadow-of-death effect might be included in within effect.

As a result of the low open-up rate and high closure rate in the 1990s, the output share of establishments set up in the previous six years declined from 1988 to 1995 (Panel A of Figure 4.4) and the share of establishments closing in the next ten years drastically increased from 1985 to 1993 (Panel B of Figure 4.4). These changes of age structure of manufacturing establishments might have reduced within effect.

Using our data, we evaluate the magnitude of learning effects and shadow-of-death effects in Japan's manufacturing. First, we estimate how quickly young establishments improve their productivity in comparison with older establishments. We estimated the following equation.

$$\ln TFP_{t,f}^{*} = \alpha^{L} + \sum_{T=0}^{16} \beta^{Ls}{}_{T} D^{Ls}_{f,T} + \sum_{T=0}^{16} \beta^{Lm}{}_{T} D^{Lm}_{f,T} + \sum_{i=1}^{49} \chi^{L}{}_{i} D^{L}{}_{f,i} + \sum_{y=1}^{Y} \delta^{L}{}_{y} D^{L}{}_{t,year} + \varepsilon^{L}{}_{t,f} \tag{4.1}$$

where an age dummy variable $D^{Lm}{}_{f,T}$ takes value one if the establishment f belongs to a firm operating multiple establishments (we call such an establishment a multi-plant establishment) and is started up T years ago. Otherwise $D^{Lm}{}_{f,T}$ takes value zero. $D^{Ls}{}_{f,T}$ takes value one if the establishment f belongs to a firm not operating multiple establishments (we call such an establishment a single-plant establishment) and is started up T years ago. Otherwise $D^{Ls}{}_{f,T}$ takes the value zero. The set of $D^{L}{}_{f,t}$ denote year dummies and the set of $D^{L}{}_{t,year}$ denote industry dummies. $TFP_{t,f}$* denotes a gap between establishment f's TFP level and simple average TFP level of that industry in year t;

$$\ln TFP_{f,t}^{*} = (\ln Q_{f,t} - \overline{\ln Q_{t}}) - \sum_{i=1}^{n} \frac{1}{2}(S_{i,f,t} + \overline{S_{i,t}})(\ln X_{i,f,t} - \overline{\ln X_{i,t}})$$

Since we found large differences of both TFP level and learning effects between single-plant establishments and multi-plant establishments, we assumed a different age coefficient between the two groups.

If we include establishments closed later within our sample, our age coefficients pick up not only learning effects but also selection effects; the effects caused by closure of establishments. As Table 4.2 shows, there is a severe selection process. Less than 50% of newly opened establishments can survive in the next 10 years. In order to exclude selection effects, we exclude establishments closed before 2003 from our observations for estimation. In the case of establishments set up before 1981, we do not know their age and all the age dummies take value zero. Such establishments plus establishments older than 16 years are the base group for the estimation. The data is pooled and we used OLS regression.

Figure 4.5 shows the estimated learning effect. Vertical line segments denote 5% confidence intervals. Multi-plant establishments and single-plant establishments have similar levels of TFP in the open-up year. But multi-plant establishments more quickly improve their productivity after the open-up.

In a similar way, we estimate the following equation in order to evaluate shadow-of-death effects.

$$\ln TFP_{t,f}{}^{*} = \alpha^{C} + \sum_{T=0}^{20} \beta^{Cs}{}_{T} D^{Cs}_{f,T} + \sum_{T=0}^{20} \beta^{Cm}{}_{T} D^{Cm}_{f,T} + \sum_{i=1}^{49} \chi^{C}{}_{i} D^{C}{}_{f,i} + \sum_{y=1}^{Y} \delta^{C}{}_{y} D^{C}{}_{t,year} + \varepsilon^{C}{}_{t,f} \tag{4.2}$$

where an closure year dummy variable $D^{Cm}{}_{f,T}$ takes value one if the establishment f is a multi-plant establishment and is closed T years later. Otherwise $D^{Cm}{}_{f,T}$ takes value zero. $D^{Cs}{}_{f,T}$ takes value one if the establishment f is a single-plant and is closed T years later. We exclude establishments, which entered after 1982 and survived until 2003, from our observations for estimation since we do not know whether they can survive for more than 21 years. Establishments which survived for more than 21 years are the base group for the estimation.

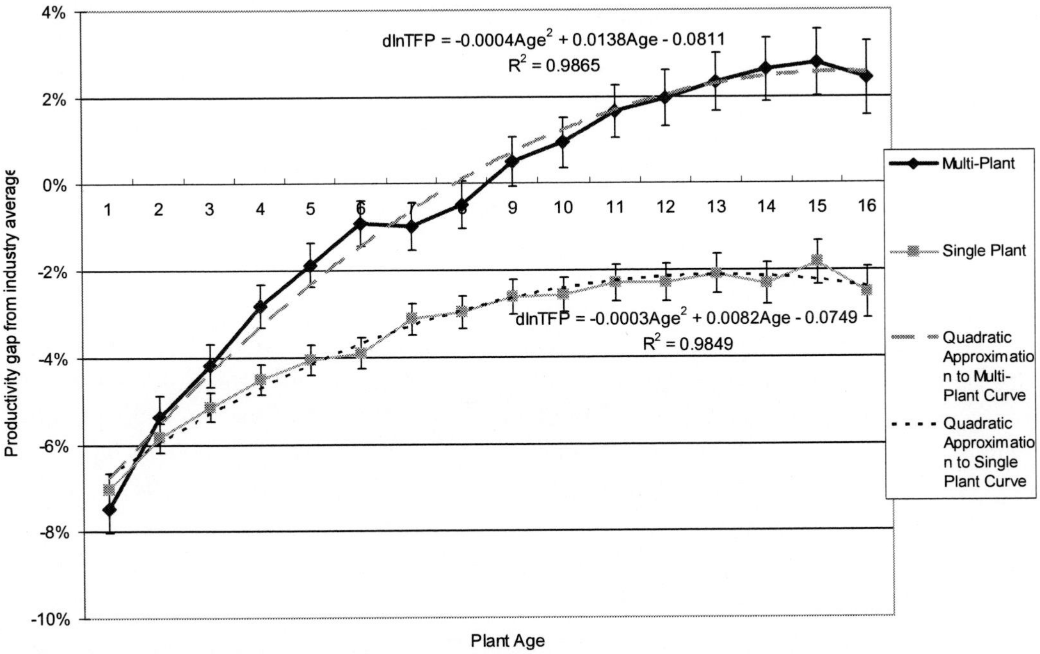

1. Base Group is the Plants which were established in 1981 or earlier and have been operating until 2003.
2. Observations of the plants which were closed before 2003 are not included in the estimations.

Figure 4.5. Age Effect on TFP by Multi and Single Plant Group.

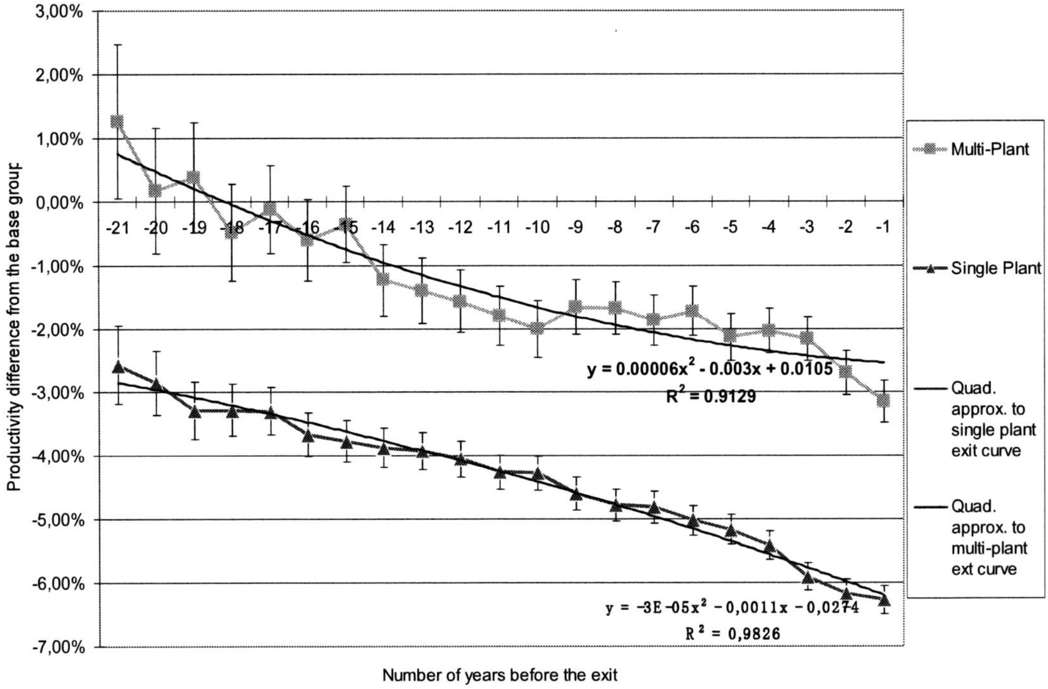

Figure 4.6. Shadow-of-Death Effect on TFP by Multi and Single Plant Group.

Figure 4.6 shows estimated shadow-of-death effect. Vertical line segments denote 5% confidence intervals. Although the average TFP level of multi-plant establishments is much higher than that of single-plant establishments, they deteriorate before the closure in a similar way. We also calculated inter-temporal changes of ratio of the average plant size of establishments set up in the same year to the industry average plant size (Table 4.3) and inter-temporal changes of ratio of the average plant size of establishments closed in the same year to the industry average plant size (Table 4.4). These tables show that behind learning effects and shadow-of-death effects on TFP, size of establishments also sharply changes overtime.

We found substantial learning effects, which works for longer than five years after the open-up, and shadow-of-death effects, which works more than five years before the closure. We also found that as a result of low open-up rates and high closure rates in the 1990s, the output share of young establishments has declined and the share of establishments closing in the next ten years has drastically increased from the 1980s to the 1990s. Therefore it seems that we can explain the decline of within effects in the period of 1990-2003, at least partly, by the low open-up rate and high closure rate in the 1990s.

5. Conclusion

Using the JIP 2006 Database and establishment-level data of the *Census of Manufactures*, we examined why Japan's TFP growth slowed down in the 1990s. The major results obtained through our analysis are follows:

1. Using JIP 2006 data, we examined Japan's TFP growth decline at the macro and the sectoral level. Like previous studies, we found that the growth rate of TFP declined at the macro level and the slowdown in TFP growth was more severe in the manufacturing than in the non-manufacturing sector.
2. Our productivity decomposition analysis based on establishment-level data showed that the decline in aggregate productivity growth was not due to a decline in the reallocation effect but due to (a) a slowdown in TFP growth within establishments; and (b) a worsening negative exit effect, meaning that closures of productive establishments were more frequent than closures of unproductive establishment.
3. In all periods and all countries, the net entry effect is positive, suggesting that productivity grows as less productive establishments exit and more productive ones enter. In addition, we find that in almost all countries, the net entry effect is the major source of overall productivity growth during recessionary periods. Yet, although Japan's recession lasted longer than those in other countries, we find that the contribution of the net entry effect is smallest, indicating that the mechanism of "creative destruction" in Japan's manufacturing sector is very weak. However, on the bright side, the contribution of the entry effect has increased in recent years.
4. Transition matrices reveal a considerable persistence in the level of productivity. 56% of establishments which originally ranked in the bottom three deciles in 1990 remained in the same three deciles in 2003. Similarly, 55% of establishments which originally ranked in the top three deciles in 1990 remained in the same three deciles in 2003. However, the survival rate is not high even in the case of establishments in the top labor-productivity groups.
5. The average productivity of establishments that exited was higher than the average productivity in that industry. Probably, this is a consequence of the relocation of production from Japan to foreign countries.
6. Without doubt, most of the decline in productivity growth comes from the decrease in the within effect. It seems also certain that the decline in the within effect was caused by the low start-up rate and high closure rate during the 1990s. This implies that an active turnover in plants is essential to boost Japan's aggregate productivity.

APPENDIX A. DEFINITION OF VARIABLES USED IN THE ECONOMETRIC ANALYSIS AND DATA SOURCES

The data source of this study is the *Census of Manufactures*, which is conducted annually by the Ministry of Economy, Trade and Industry (METI). The survey covers all manufacturing plants with four or more employees, and it excludes plants that ceased operating or whose employment fell below the survey's threshold of four employees. It contains detailed information on plant characteristics, such as output, employees, intermediate inputs, tangible capital, location, etc. The available data cover the period 1981-2003.

A1. Construction of Variables to Measure Productivity

Output

Gross output is measured as the sum of shipments, revenues from repairing and fixing services, and revenues from performing subcontracted work. Gross output is deflated by output deflators derived from the JIP 2006. Real value added is defined as real gross output minus real intermediate inputs.

Intermediate Inputs

Intermediate inputs are defined as the sum of raw materials, fuel, electricity and subcontracting expenses for consigned production used by the plant. Intermediate inputs are deflated by intermediate input deflators provided in the JIP 2006.

Labor input

As labor input, we used each firm's total number of workers multiplied by the sectoral working hours from the JIP 2006.

Capital Stock

Using the nominal book values of tangible fixed assets including buildings, machinery, tools and transport equipment, we calculated the net capital stock of plant *i* in industry *j* in constant 1995 prices as follows:

$$K_{it} = BV_{it} * (INK_{jt} / IBV_{jt})$$

where BV_{it} represents the book value of firm *i*'s tangible fixed capital in year *t*, INK_{jt} stands for the net capital stock of industry *j* in constant 1995 prices, and IBV_{jt} denotes the book value of industry *j*'s capital. INK_{jt} is calculated as follows. First, as a benchmark, we took the data on the book value of tangible fixed assets in 1976 from the *Census of Manufactures 1976* published by METI. We then converted the book value of year 1976 into the real value in constant 1995 prices using the net fixed assets deflator provided in the *Annual Report on National Accounts* published by the Cabinet Office, Government of Japan. Second, the net capital stock of industry *j*, INK_{jt}, for succeeding years was calculated using the perpetual inventory method. We used the investment deflator in the JIP 2006. The sectoral depreciation rate used is also taken from the JIP 2006.

Cost Shares

Labor costs are defined as total salaries and intermediate costs are defined as the sum of raw materials, fuel, electricity and subcontracting expenses for consigned production provided in the *Census of Manufactures*, respectively. Capital costs were calculated by multiplying the real net capital stock with the user cost of capital. The latter was estimated as follows:

$$c_k = \frac{1-z}{1-u} p_k \{(1-u)i + \delta_j - (\frac{\dot{p}_k}{p_k})\}$$

where p_k, i, δ, u and z are the price of investment goods, the interest rate, the depreciation rate, the corporate tax rate, and the present value of depreciation deduction on a unit of nominal investment, respectively. Data on investment goods prices, interest rates, and corporate tax rates were taken from the JIP 2006, the *Bank of Japan* and the *Ministry of Finance Statistics Monthly*. The depreciation rate for each sector is taken from the JIP 2006. We measure the cost share of each factor by dividing the costs of each factor by total costs, which is the sum of labor costs, intermediate input costs, and capital costs.

A2. Treatment of Outliers

Data for the Analysis of Labor Productivity

To begin with, we excluded plants that reported either zero employees or a negative number of employees, and negative value added. Next, we defined as outliers those plants whose annual average labor productivity fell outside four times standard deviations of labor productivity and eliminated these plants from our sample.

Data for the Analysis of TFP

We dropped plants that did not provide information on the book value of tangible fixed asset and plants that report zero total salaries and intermediate inputs. We also excluded outliers from our sample using the same rule as that applied for labor productivity.

REFERENCES

Ahn, S., K. Fukao and H. U. Kwon (2004) "The Internationalization and Performance of Korean and Japanese Firms: An Empirical Analysis Based on Micro Data" *Seoul Journal of Economics*, Vol. 17, No. 4, pp. 439-82.

Aw, B. Y., X. Chen and M. J. Roberts (2001) "Firm-level Evidence on Productivity Differentials and Turnover in Taiwanese Manufacturing," *Journal of Development Economics*, Vol. 66, No.1, pp. 51-86.

Baily, M. N., C. Hulten and D. Campbell (1992) "Productivity Dynamics in Manufacturing Plants," *Brookings Papers on Economics Activity: Microeconomics*, Vol. 2, pp. 187-249.

Baldwin, J. R. and W. Gu (2006) "Plant Turnover and Productivity Growth in Canadian Manufacturing," *Industrial and Corporate Change*, Vol.15, pp. 417-465.

Bartelsman, E. J. and M. Doms (2000) "Understanding Productivity: Lessons from Longitudinal Microdata," *Journal of Economic Literature*, Vol. 38, No.3, pp. 569-594.

Bartelsman, E. J., J. Haltiwanger and S. Scarpetta (2004) "Microeconomic Evidence of Creative Destruction in Industrial and Developing Countries," *Policy Research Working Paper Series* 3464, The World Bank.

Disney, R., J. Haskel, and Y. Heden (2003) "Restructuring and Productivity Growth in UK Manufacturing," *Economic Journal*, Vol.113, pp.666-694.

Doms, M., J,B, Jensen, F. Kramarz, K. Motohashi, and V. Nocke (1995) "A Micro Economic Comparison of the Manufacturing Sectors in France, Japan and the United States," paper presented at the NBER Summer Institute, Productivity Workshop, July 1995, Cambridge. MA.

Farrell, D., M. Baily and J. Remes (2005) "US Productivity After the Dot Com Bust," MGI Report "U.S. Productivity Growth, 1995-2000" at <http://www.mckinsey.com/mgi/publication/us/index.asp>

Foster, L., J. Haltiwanger and C. J. Krizan (2001) "Aggregate Productivity Growth: Lessons from Microeconomic Evidence," in C. R. Hulten, E.R. Dean, and M. J. Harper (eds.), *New Contributions to Productivity Analysis*, Chicago: The University of Chicago Press, pp. 303-372.

Fukao, K., T. Inui, H. Kawai and T. Miyagawa (2004) "Sectoral Productivity and Economic Growth in Japan, 1970-98; An Empirical Analysis Based on the JIP Database," in T. Ito and A. K. Rose (eds.), *Growth and Productivity in East Asia*, National Bureau of Economic Research-East Asia Seminar on Economics, Vol. 13, Chicago: University of Chicago Press, pp.177-227.

Fukao, K. and H.U. Kwon (2006) "Why Did Japan's TFP Growth Slow Down in the Lost Decade? An Empirical Analysis Based on Firm-Level Data of Manufacturing Firms," *The Japanese Economic Review*, Vol. 57, No. 2, pp. 195-227.

Fukao, K., S. Hamagata, T. Inui, K. Ito, H.U. Kwon, T. Makino, T. Miyagawa, Y. Nakanishi, and J. Tokui (2006) "Estimation Procedures and TFP Analysis of the JIP Database 2006 Provisional Version," paper presented at the EU KLEMS 3rd Consortium Meeting, May 7-9, 2006, Valencia (Spain).

Good, D. H., M. I. Nadiri and R. C. Sickles (1997) "Index Number and Factor Demand Approaches to the Estimation of Productivity," in M.H. Pesaran and P. Schmidt (eds.), *Handbook of Applied Econometrics: Vol. 2. Microeconometrics*, Oxford, England: Basil Blackwell, pp. 14-80.

Griliches, Z. and H. Regev (1995) "Productivity and Firm Turnover in Israeli Industry: 1979-1988," *Journal of Econometrics*, Vol. 65, No.1, pp.175-203.

Hayashi, F. and E. C. Prescott (2002) "The 1990s in Japan: A Lost Decade," *Review of Economic Dynamics*, Vol. 5, No. 1, pp. 206-35.

Jorgenson, D. W. and K. Motohashi (2003) "The Role of Information Technology in the Economy: Comparison between Japan and the United States," prepared for *RIETI/KEIO Conference on Japanese Economy: Leading East Asia in the 21st Century?* Keio University.

Kawamoto, T. (2004) "What Do the Purified Solow Residuals Tell Us about Japan's Lost Decade?" Bank of Japan IMES Discussion Paper Series, No. 2004-E-5, Bank of Japan, Tokyo.

Kiyota, K. and M. Takizawa (2006) "The Shadow of Death: Pre-exit Performance of Firms in Japan," mimeo.

Motohashi, K. (2001) "Development of Longitudinal Micro-Datasets and Policy Analysis for Japanese Industrial Sectors," *RIETI Discussion Paper Series* 01-E-007, RIETI, Tokyo.

New Business Creation Subcommittee, New Growth Policy Committee, the Industrial Structure Council (2002) *New Business Creation Subcommittee Report: For the Promotion of New Business and the Facilitation of Growth*, Ministry of Economy, Trade and Industry, Government of Japan.

Nishimura, K. G., T. Nakajima and K. Kiyota (2005) "Does the Natural Selection Mechanism Still Work in Severe Recessions? Examination of the Japanese Economy in the 1990s." *Journal of Economic Behavior and Organization*, forthcoming.

Shimizu, M. and K. Miyagawa (2003) *Sannyu-Taishutsu to Takakukano Keizai Bunseki* (Economic Analysis on Entry-Exit and Diversification of Business), Tokyo: Keio University Press.

Shinbo, K., M. Takahashi, and M. Omori (2005) "Kogyo Tokei Panel Data no Sakusei: Sangyo Kozo Database no Ikkan Toshite (Creation of the Census of Manufactures Panel Data: As Part of an Industrial Structure Database)," *RIETI Policy Discussion Paper Series*. 05-P-001, RIETI, Tokyo.

Small and Medium Enterprise Agency, Ministry of Industry, Trade and Industry, Japanese Government (2001) *2001 White Paper on Small and Medium Enterprises in Japan*, Ministry of Industry, Trade and Industry, Japanese Government, Tokyo (available at http://www.chusho.meti.go.jp /sme_english/index.html).

Small Business Administration, U.S. Government (1998) *The State of Small Business: A Report of the President*, Small Business Administration, U.S. Government, Washington D.C..

Study Group on "Industry Hollowing-out" and Tariff Policy, Ministry of Finance, Japanese Government (2002) *Chairperson's Report*, Ministry of Finance, Japanese Government, Tokyo.

In: Micro-Evidence for the Dynamics of Industrial Evolution ISBN: 978-1-60456-388-7
Editors: A. Heshmati, J.-D. Lee

Chapter 4

THE IMPACT OF INTERNATIONALIZATION ON EMPLOYMENT AND PRODUCTIVITY GROWTH: A COMPARATIVE ANALYSIS OF JAPANESE AND KOREAN MICRO-DATA

Sanghoon Ahn[1,*]
Korea Development Institute, P. O. Box 113, Cheongnyang, Seoul 130-012, Korea
Keiko Ito[2,†]
Senshu University, 2-1-1, Higashi-mita, Tama-ku, Kawasaki, Kanagawa 214-8580, Japan

ABSTRACT

This study examines the economic impact of outbound FDI and trade by comparing Korea and Japan by the means of regression analyses. The comparison of these two countries is motivated by the fact that: (i) Japan has a longer experience in outbound FDI and overseas production than Korea, (ii) Korea's outbound FDI is much more concentrated in China than Japan's, (iii) despite high global competitiveness of Korean firms in a number of areas, there still exists a significant gap between Korea and Japan in exporting firms productivity levels. The closer similarity in skills of workers in Korea and China suggest that Korean workers are much more strongly affected by competition from Chinese workers than their Japanese counterpart. This study reconfirms and provides a more nuanced perspective on the previous finding of negative correlation between outbound FDI and growth in employment and productivity in manufacturing in Korea and Japan. However, the negative effect of outsourcing is more pronounced in the case of Japan with more advanced outsourcing activities than Korea.

* Tel.: +82-2-958-4107, Fax.: +82-2-960-4570, e-mail:ahn@kdi.re.kr
† Tel.: +81-44-900-7818, Fax.: +81-44-911-0467, e-mail:keiko-i@isc.senshu-u.ac.jp

1. INTRODUCTION

During the past decade, the economies of East Asia have been becoming ever more interdependent. A key role in this trend is played by China and its contribution to the international division of labor in the region. Japanese multinational enterprises (MNEs) in the manufacturing sector have been investing in the country since the early 1990s in the wake of an earlier investment spree in Southeast Asia. In 1990, China's share in Japanese outward foreign direct investment (FDI) accounted for only 1.0% of the total in terms of the overall total value of FDI in the manufacturing sector. However, by 2002, the corresponding figure was 9.6%, rising further to 27.5% in 2004. In terms of Japan's FDI cases in the manufacturing sector, China's share accounted for 7.4% in 1990, going up to 35.4% in 2002 and to 46.8% in 2004.[1] In the case of Korea, outbound FDI overall has been growing since the mid-1980s, experiencing a rapid increase during the years 1993-96 and another spurt since the end of the 1990s. In Korea's case, too, there has been a sharp increase in the amount of FDI to China since 1998, notably by small and medium-sized enterprises.[2] China is now the most important destination of manufacturing FDI for both Korea and Japan. Looking at Chinese statistics on investment flows and cumulative inward investment amounts in all industries, Korea and Japan have been the top-two investors in China in recent years in terms of investment amounts if Hong Kong is excluded (Table 1 (a)).

The rapid increase in outbound FDI to China has also contributed to the increase in trade with China. Moreover, in an argument that has similarly been heard in many other developed countries, some claim that Korean and Japanese firms have been shutting down factories at home and shifting employment abroad to countries with cheaper labor (particularly China). According to the *Census on Basic Characteristics of Establishments* compiled by the National Statistical Office of Korea, the total number of persons employed in manufacturing in that country decreased by 12.0% (from 3.9 million workers to 3.4 million workers) during the period from 1993-2004 (an annual decrease of 1.1%), while the total number of manufacturing establishments increased by 16.6% (from 281,590 to 328,338) during the same period (an annual increase of 1.5%). In the case of Japan, between 1996 and 2001, the total number of establishments and the number of persons engaged in manufacturing decreased by 15.6% (or 3.1% annually) from 771,906 to 651,111 and by 13.9% (or 2.8% annually) from 12.9 million to 11.1 million workers, respectively (Statistics Bureau, various years).

---

[1] Japan's gross FDI figures are taken from Ministry of Finance, Japan (2006).

[2] See, among others, Ahn *et al.* (2005).

Table 1. Inward FDI in China, by source country

(a) Number of inward FDI projects and amount of investment (Unit: US$10,000)

	No. of projects				Amount of investments fulfilled			
	2003		2004		2003		2004	
World Total	41,081	(100.0)	43,664	(100.0)	5,350,467	(100.0)	6,062,998	(100.0)
Hong Kong	13,633	(33.2)	14,719	(33.7)	1,770,010	(33.1)	1,899,830	(31.3)
Japan	3,254	(7.9)	3,454	(7.9)	505,419	(9.4)	545,157	(9.0)
Taiwan	4,495	(10.9)	4,002	(9.2)	337,724	(6.3)	311,749	(5.1)
Macau	580	(1.4)	715	(1.6)	41,660	(0.8)	54,639	(0.9)
Korea	4,920	(12.0)	5,625	(12.9)	448,854	(8.4)	624,786	(10.3)
U.S.A.	4,060	(9.9)	3,925	(9.0)	419,851	(7.8)	394,095	(6.5)
Canada	901	(2.2)	995	(2.3)	56,351	(1.1)	61,387	(1.0)
Europe	2,074	(5.0)	2,423	(5.5)	393,031	(7.3)	423,904	(7.0)
Germany	451	(1.1)	608	(1.4)	85,697	(1.6)	105,848	(1.7)
France	269	(0.7)	289	(0.7)	60,431	(1.1)	65,674	(1.1)
Italy	297	(0.7)	358	(0.8)	31,670	(0.6)	28,082	(0.5)
Netherland	189	(0.5)	199	(0.5)	72,549	(1.4)	81,056	(1.3)
U.K.	438	(1.1)	488	(1.1)	74,247	(1.4)	79,282	(1.3)
ASEAN-5	2,128	(5.2)	2,156	(4.9)	285,309	(5.3)	290,962	(4.8)
Singapore	1,144	(2.8)	1,279	(2.9)	205,840	(3.8)	200,814	(3.3)
Indonesia	143	(0.3)	122	(0.3)	15,013	(0.3)	10,452	(0.2)
Malaysia	350	(0.9)	352	(0.8)	25,103	(0.5)	38,504	(0.6)
Philippines	297	(0.7)	241	(0.6)	22,001	(0.4)	23,324	(0.4)
Thailand	194	(0.5)	162	(0.4)	17,352	(0.3)	17,868	(0.3)
Others	5,036	(12.3)	5,650	(12.9)	1,092,258	(20.4)	1,456,489	(24.0)
Br. Virgin Iss.	2,218	(5.4)	2,641	(6.0)	577,696	(10.8)	673,030	(11.1)

Figures in parentheses indicate World Total in percent.
Source: Ministry of Commerce, People's Republic of China (2004, 2005).

Table 1. (Continued)

(b) Cumulative number and amount of investment of inward FDI projects								(Unit: US$100 million)
	No. of projects				Amount of investments fulfilled			
	up to 2003		up to 2004		up to 2003		up to 2004	
World Total	465,277	(100.0)	508,941	(100.0)	5,015	(100.0)	5,612	(100.0)
Hong Kong	224,509	(48.3)	239,228	(47.0)	2,226	(44.4)	2,416	(43.0)
Japan	28,401	(6.1)	31,855	(6.3)	414	(8.3)	468	(8.3)
Taiwan	60,186	(12.9)	64,188	(12.6)	365	(7.3)	396	(7.1)
Macau	8,407	(1.8)	9,122	(1.8)	52	(1.0)	57	(1.0)
Korea	27,128	(5.8)	32,753	(6.4)	197	(3.9)	259	(4.6)
U.S.A.	41,340	(8.9)	45,265	(8.9)	441	(8.8)	480	(8.6)
Canada	6,941	(1.5)	7,936	(1.6)	39	(0.8)	45	(0.8)
Europe	16,158	(3.5)	18,581	(3.7)	379	(7.6)	421	(7.5)
Germany	3,504	(0.8)	4,112	(0.8)	89	(1.8)	99	(1.8)
France	2,302	(0.5)	2,591	(0.5)	61	(1.2)	68	(1.2)
Italy	2,137	(0.5)	2,495	(0.5)	25	(0.5)	28	(0.5)
Netherland	1,254	(0.3)	1,453	(0.3)	51	(1.0)	59	(1.0)
U.K.	3,856	(0.8)	4,344	(0.9)	114	(2.3)	122	(2.2)
ASEAN-5	21,158	(4.5)	23,314	(4.6)	321	(6.4)	350	(6.2)
Singapore	11,871	(2.6)	13,150	(2.6)	235	(4.7)	255	(4.6)
Indonesia	1,079	(0.2)	1,201	(0.2)	13	(0.3)	14	(0.2)
Malaysia	2,888	(0.6)	3,240	(0.6)	31	(0.6)	35	(0.6)
Philippines	1,945	(0.4)	2,186	(0.4)	16	(0.3)	19	(0.3)
Thailand	3,375	(0.7)	3,537	(0.7)	25	(0.5)	27	(0.5)
Others	31,049	(6.7)	36,699	(7.2)	582	(11.6)	718	(12.8)
Br. Virgin Iss.	8,877	(1.9)	11,518	(2.3)	302	(6.0)	369	(6.6)

Figures in parentheses indicate World Total in percent.
Source: Ministry of Commerce, People's Republic of China (2004, 2005).

Looking at average firm size and productivity for the period from 1994 to 2002 for Japan (Figure 1A), average firm size has been decreasing in all industry categories except the low-tech industries, while only high-tech industries appear to have experienced an improvement in the average TFP level.[3] A similar pattern, although not quite as pronounced, of declining average plant size and relatively sluggish productivity growth in low-tech industries is observed in Korean manufacturing (Figure 1B).

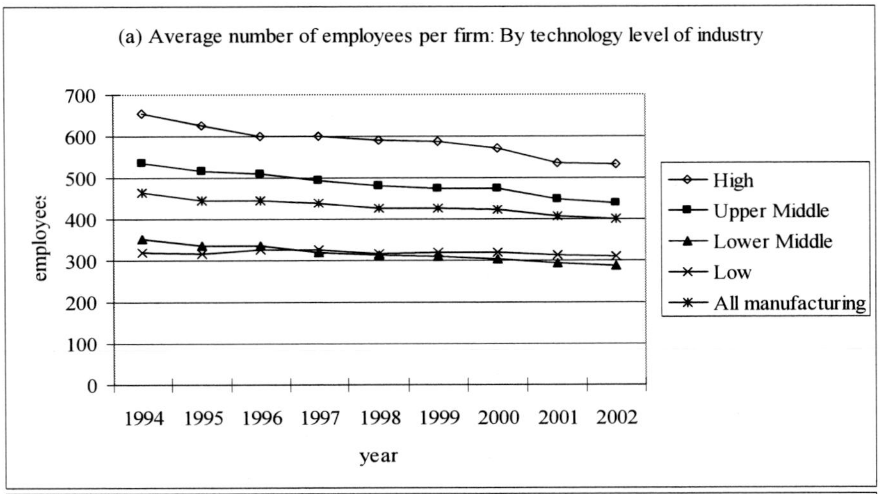

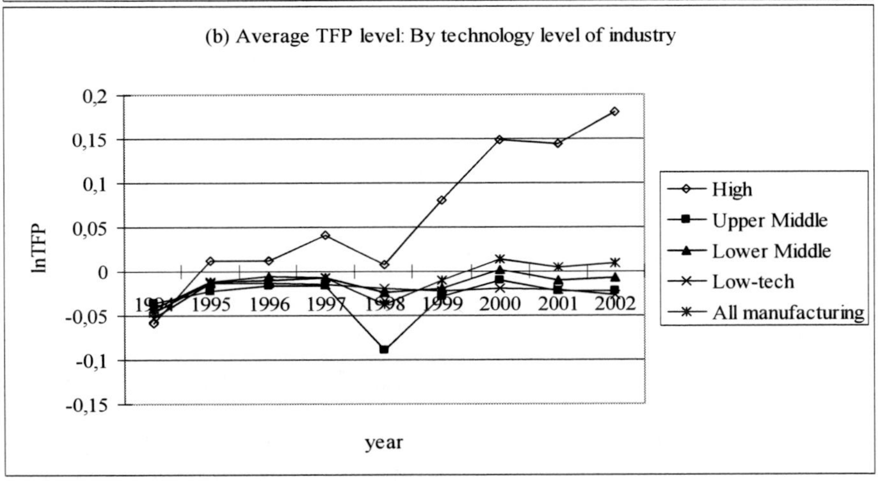

Source: Authors' calculation based on the firm-level data underlying the Basic Survey of Japanese Business Structure and Activities.

Figure 1A. Average number of employees per firm and TFP level for Japan–By technology level of industry.

---

3 Average firm size and productivity are calculated using the firm-level data underlying the *Basic Survey of Business Structure and Activities* annually conducted by the Ministry of Economy, Trade and Industry (METI). The classification of manufacturing industries is based on the OECD's (1997) classification (see, Appendix). A detailed data description is provided in Section 3 of this paper.

2 According to the World Bank's classification, China was as a low-income country until 1998 and has been a lower-middle income country since 1999.

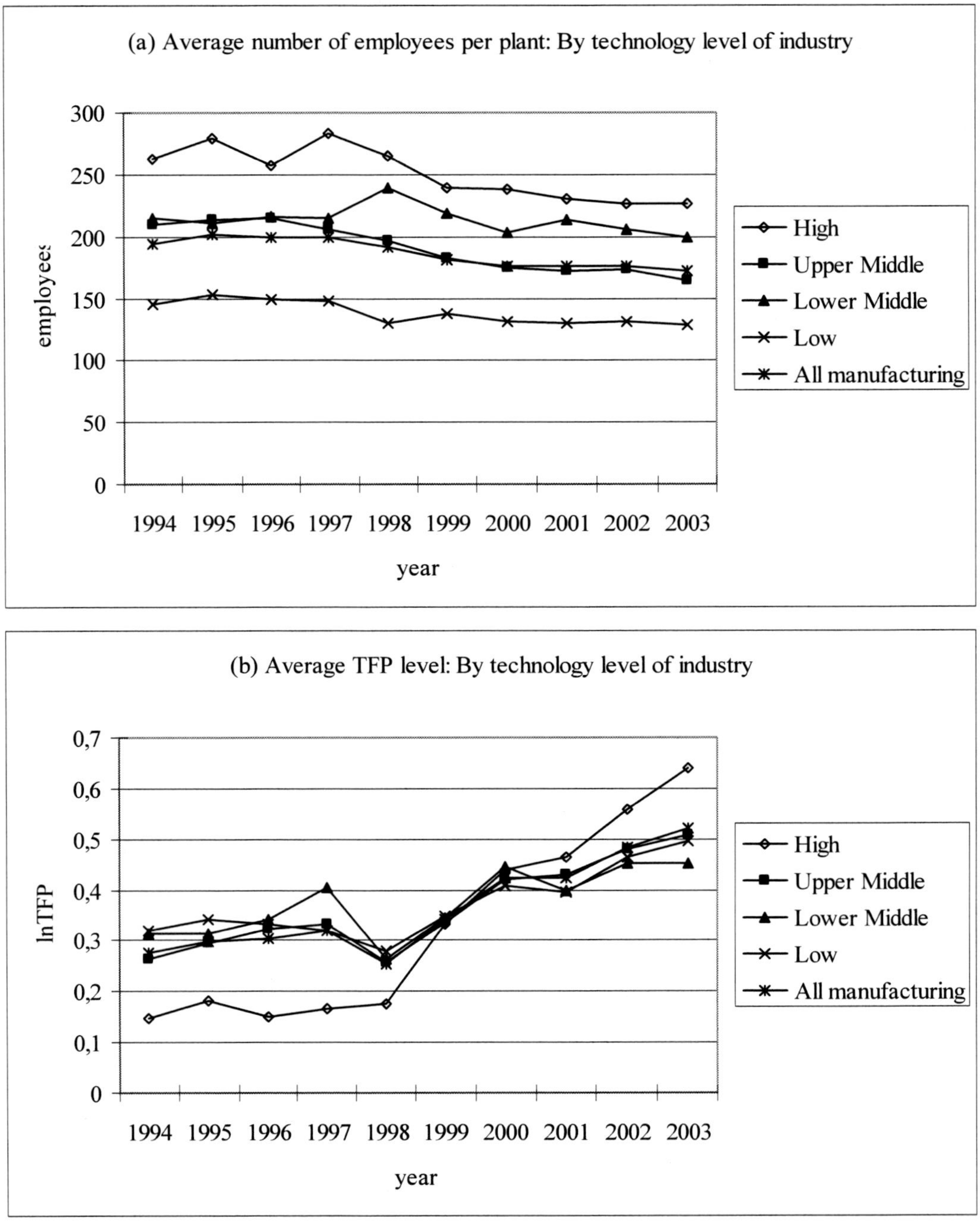

Source: Authors' calculation based on the plant-level data underlying the Annual Report on Mining and Manufacturing Survey.

Figure 1B. Average number of employees per plant and TFP level for Korea –By technology level of industry.

These statistics suggest a sort of "substitutability" between the demand for domestic and foreign labor and appear to lend support to the view that globalization leads to a "hollowing-out" effect. In the United States and Europe, for example, numerous studies have examined the link between home labor markets and globalization focusing on issues such as outsourcing, foreign production, and import competition from low-wage countries, and some, such as Brainard and Riker (1997, 2001) and Muendler and Becker (2006) have shown that employment abroad indeed seems to substitute for employment at home. In addition, Harrison and McMillan (2006), for instance, found that the sign of the correlation between home employment and employment abroad depended upon the crucial distinction between affiliates in high-income and low-income countries, and that U.S. employment and employment in low-income (high-income) countries were substitutes (complements). Although their analysis is based on firm-level data and focuses on the allocation of employment at home and abroad within a firm, their results are consistent with evidence on trade and employment linkages presented by Bernard, Jensen, and Schott (2006). The latter, using U.S. plant-level panel data, analyzed the impact of imports from low-wage countries on plant survival, employment, industry switching, etc., and found that industry-wide imports harm U.S. manufacturing employment at the plant level when those imports are from low wage countries.

In addition, globalization possibly affects productivity at home. For example, it has been suggested that productivity may be negatively affected if the contraction of home activities following relocation reduces efficiency through decreasing plant-level scale effects (Barba Navaretti and Venables, 2004). Moreover, the "hollowing-out" effect may be exacerbated if relocation abroad further reduces domestic efficiency through decreasing regional or industry-level agglomeration effects. At the same time, however, the international division of labor may also help to boost the productivity of the primary factors of production, because it allows firms to specialize in activities they perform relatively well or to realize sizeable cost-savings by utilizing imported intermediate inputs. A number of studies have attempted to assess the impact of foreign production or trade on productivity at home. Amiti and Wei (2006), for instance, relying on U.S. industry-level data (96 manufacturing industries), found that both material and service outsourcing had a positive impact on industry-level total factor productivity (TFP). Meanwhile, Görg, Hanley, and Strobl (2005) analyzed this issue using plant-level data for Irish manufacturing and found that outsourcing benefits plant productivity.

In the case of Japan, Head and Ries (2002) analyzed the impact of offshore production on skill intensity at home using firm-level data for Japanese MNEs, while Ito and Fukao (2005a, 2005b) analyzed to what extent offshoring contributed to changes in the skill structure of employment at home using industry-level data. Ito (2005) showed that industries exposed to fierce competition from cheap imports from low- and lower-middle income countries displayed slower sales and employment growth rates than other industries. On the other hand, Kimura and Kiyota (2006) found that firms that engage in outbound FDI exhibit productivity growth that is on average 1.8 percentage-points higher than firms that do not. However, another study along these lines by Hijzen, Inui, and Todo (2006) did not detect a statistically significant positive effect of FDI on productivity. Thus, various effects of globalization on firms' performance have been examined. However, preceding studies have sometimes produced mixed results.

In the case of Korea, the economic impact of outbound FDI and trade on industry performance as measured by employment or productivity is an area that remains under-explored, although here, too, concerns about "hollowing out" are increasing. One of the few studies on this issue in Korea is Ahn *et al.* (2005), which found that: (1) an increase in the growth rate of outbound FDI in a certain industry leads to an average increase in the employment growth and TFP growth of establishments in that industry; (2) the higher the share of outward FDI to China in a certain industry, the lower the average employment growth and TFP growth of establishments in that industry; and, (3) the greater the share of outward FDI to the U.S. in a certain industry, the larger the positive impact on the TFP growth of establishments in that industry.

The results of Ahn *et al.* (2005) suggest that the sign of the economic impact of FDI and trade on industry performance depends on the crucial distinction between high- and low-income FDI destinations or trade partners, which is in line with the results of the studies by Harrison and McMillan (2006) and Bernard, Jensen, and Schott (2006) referred to above. Thus, the empirical evidence suggests that a key factor determining the sign of the impact of FDI and trade on the home economy are the wage and technological levels of the partner country. Given the fact that China — still a low-wage country[2]— has been gaining a larger and larger share in both Korea's and Japan's outbound FDI and trade, China's impact on home industry performance in these two countries is an issue that deserves further investigation.

The purpose of this paper is to examine the economic impact of outbound FDI and trade by comparing the cases of Korea and Japan through a series of regression analyses. The comparison of these two countries is potentially instructive for the following reasons. First, Japan has a relatively longer experience in outbound FDI and overseas production than Korea, and the economic impact of globalization may possibly differ as a result of the difference in their overseas business experience. Second, Korea's outbound FDI is much more concentrated in China than Japan's. Moreover, the figures in Table 1 on Korean and Japanese investment flows to and cumulative investment amounts in China imply that in recent years, Korea has been investing more actively in China than Japan.[3] The difference in the two country's distribution of outward FDI, i.e., the much higher concentration of Korean FDI in China, may result in differences in the impact on the home economy. Third, although Korean firms are now quite competitive in the global market in the fields of electrical machinery, electronic components, communication equipment, automobiles, etc., there nevertheless still exists a significant gap between Korea and Japan in productivity levels, especially in major exporting industries.[4] Thus, referring to the findings of Riker and Brainard (1997), who found that workers competed across affiliates but the competition was between affiliates in countries with similar workforce skill levels, we would expect that if the skill level of workforces in China is much closer to that in Korea than in Japan, Korean workers are much more strongly affected by competition from Chinese workers.

The rest of the paper unfolds as follows. The next section provides an overview of Korea's and Japan's outbound FDI and trade patterns since 1990. Covering the period of 1990-2002 and using three-digit industry-level data, we look at changes in trends and other salient features related to the outbound FDI and trade of these countries. Special care is taken to identify changes in the composition of investment and trade by region and by industry.

---

[3] In Table 1, we can see that the amount of investment fulfilled by Japan and Korea in 2003 and 2004 were comparable, but Japan's cumulative amount of investment up to 2004 was still far higher than Korea's.

[4] According to industry-level data provided by the Groningen Growth and Development Centre (http:// www.ggdc.net), the labor productivity level per hour worked is much higher in Japan than in Korea in all manufacturing industries. Using the average U.S. productivity level as a common yardstick, Japan's average productivity stands at approximately 70% and Korea's at around 30% of the U.S. level. In exporting industries, such as electrical machinery and motor vehicles, Korea's productivity level is only slightly more than one-third of Japan's productivity level. Also refer to Table 4.5 in Ahn *et al.* (2004).

Section 3 provides a brief survey of the literature and then presents the regression analysis on the economic impact of outbound FDI and trade in Korea and Japan. The focus is the impact on TFP growth and employment growth in the manufacturing sector and special attention is paid to identifying the impact from transactions with China, the U. S., Korea and Japan. In particular, we attempt to distinguish the impact of transactions with China – a low-wage country – from the impact of transactions with other countries. The final section summarizes the findings of the paper.

2. Recent Patterns in Outbound FDI and Trade in Korea and Japan

Outward FDI by South Korean firms has grown steadily since the mid-1980s, particularly due to liberalization policies first adopted at that time. Korean outbound FDI rapidly increased during the years 1993-96 but then slowed down somewhat afterwards. The share of the total FDI stock in GDP increased from 0.2% in 1980 to 5.7% in 2003, which, however, is still well below that of the U.S. (18.8%) and most other developed countries (26.4% on average).[5] Thus, although Korea has been a late starter in terms of conducting investments abroad, the importance of outbound FDI has been increasing rapidly.

In the case of Japan, the history of outbound FDI dates back to before World War I, when a number of trading and insurance companies set up abroad (Paprzycki and Fukao, 2005). Outbound FDI started to increase in the 1960s as major Japanese manufacturers established production facilities in various developing countries in response to import substitution policies in these countries. Outbound FDI increased further when in the 1980s, several Japanese automobile manufacturers shifted production to the United States in order to mitigate trade friction. However, it was only in the wake of the rapid appreciation of the yen following the Plaza Accord in 1985 that FDI by Japanese manufacturing firms truly took off. Much of the FDI during this period went to Southeast Asia, a trend that continued until the late 1990s, when FDI in China gained greater prominence.

Tables 2A and 2B show the shares of Korea's and Japan's outbound FDI, exports and imports in the manufacturing sector by destination and level of technology for 1990 and 2002. Manufacturing industries are classified by technology-intensity following the OECD (OECD, 1997) and fall into low, medium-low, medium-high and high-technology industries.[6] Looking at the top panel of Table 2A, for example, the rows show the share of each destination in total outstanding FDI by Korean firms for a particular level of technology.

[5] These figures are based on UNCTAD's *World Investment Report*, 2004.
[6] Details of the classification are shown in the Appendix.

Table 2A. Share of each country in Korean outbound FDI, exports, and imports: By technology level

Technology Level	*Shares in Total Korean Outbound FDI by Technology Level (%)*											
	To CHINA			To JAPAN			To USA			To World		
	1990	2002	Change	1990	2002	Change	1990	2002	Change	1990	2002	Change
High	2.68	12.08	9.40	0.00	0.98	0.98	26.28	36.09	9.82	7.41	38.15	30.74
Upper Middle	2.15	34.11	31.96	0.79	1.72	0.93	11.61	12.04	0.43	26.39	25.09	-1.31
Lower Middle	1.17	39.62	38.44	0.52	0.24	-0.28	51.13	23.90	-27.23	33.82	15.40	-18.41
Low	2.66	37.17	34.51	0.20	0.13	-0.07	19.41	8.97	-10.44	32.38	21.37	-11.02
Mfg. Total	2.02	27.21	25.18	0.45	0.87	0.42	28.59	22.39	-6.20	100.00	100.00	0.00

Technology Level	*Shares in Total Korean Exports by Technology Level (%)*											
	To CHINA			To JAPAN			To USA			To World		
	1990	2002	Change	1990	2002	Change	1990	2002	Change	1990	2002	Change
High	0.17	14.13	13.96	16.72	9.05	-7.67	31.13	19.59	-11.54	4.96	11.73	6.78
Upper Middle	1.10	12.21	11.11	10.01	4.39	-5.62	35.62	30.66	-4.96	23.14	52.46	29.32
Lower Middle	0.75	17.47	16.72	22.13	21.62	-0.51	40.70	15.33	-25.38	38.61	20.24	-18.37
Low	0.58	15.41	14.84	26.54	13.96	-12.58	23.12	21.92	-1.20	33.29	15.57	-17.72
Mfg. Total	0.74	14.00	13.26	20.52	9.92	-10.61	33.20	24.90	-8.30	100.00	100.00	0.00

Technology Level	*Shares in Total Korean Imports by Technology Level (%)*											
	From CHINA			From JAPAN			From USA			From World		
	1990	2002	Change	1990	2002	Change	1990	2002	Change	1990	2002	Change
High	0.08	11.32	11.24	42.25	26.26	-15.99	22.18	14.44	-7.74	9.24	12.35	3.11
Upper Middle	0.53	8.71	8.18	43.25	34.72	-8.53	21.88	18.84	-3.04	41.06	37.61	-3.45
Lower Middle	1.90	14.37	12.46	27.76	20.58	-7.17	25.38	12.42	-12.96	29.63	28.99	-0.64
Low	3.13	26.97	23.84	10.99	5.65	-5.33	30.47	18.40	-12.07	20.07	21.05	0.98
Mfg. Total	1.42	14.52	13.10	32.09	23.46	-8.63	24.67	16.34	-8.33	100.00	100.00	0.00

Note: Highlighted cells indicate the top shares.

Sources: Statistics Canada, *World Trade Analyzer*; Export-Import Bank of Korea.

As the figures for outbound FDI to the world show, the importance of high-tech industries has increased substantially during the period from 1990 to 2002, while the importance of medium-low and low-tech industries has decreased. The second and third panels of Table 2A show the share of each trade partner in total Korean exports and imports by technology level. These figures provide further illustration of the growing importance of high-tech industries in Korea's exports, although the import share distribution across technology levels remained relatively stable during this period. As for outbound FDI, China was the most important destination for Korea in 2002 with a share of 27.2%, surpassing the United States which had held the largest share in 1990. However, in high-tech industries, the United States still received the largest share of Korean outward FDI as of 2002 (36.1%), while in other industries, China held the largest share. As for exports, although the share of the United States has decreased during this period, the world's largest economy still remains the most important destination for Korean exports. On the other hand, Japan's share in Korean exports has declined remarkably, although the island neighbor's share in Korea's medium-low tech industry exports was still the highest as of 2002. Japan has been the most important origin of Korea's manufacturing imports despite the fact that Japan's share of Korean imports has decreased across all technology levels. In contrast, China's importance as an origin of Korean imports has increased across all technology levels, but particularly so in low-tech industries. Overall, what the figures in Table 2A suggest is that China's share in Korea's outbound FDI, exports, and imports have all increased, while the share of the U.S. and Japan have both decreased in most categories. However, the United States still remains the most important destination for Korean exports, while Japan remains the most important origin of Korean imports; but, China now is the most important destination of Korean outbound FDI.

Table 2B, showing the same kind of information for Japan, suggests that the trends in FDI and trade are very similar. China's share in outward FDI, exports and imports has increased across all technology levels, while the share of the United States has decreased in most categories.[1] Korea's shares have remained relatively stable, although in Japan's trade with the country, we can see a shift from low-tech to higher-tech industries. These patterns reflect the fact that in the case of both Korea and Japan, major firms in leading export industries, such as computers, telecommunications equipment and automobiles, have recently relocated some segments of their production processes abroad and established new export bases, notably in China. However, there is also a conspicuous difference between Korea and Japan: although China's importance in the trade of the two countries has increased in both cases, the increase in Japan's imports from China is much more pronounced than the increase in its exports (Table 2B); in contrast, in the case of Korea, the change in China's shares in Korean exports and imports is almost the same (Table 2A). Moreover, the share of high-tech and medium high-tech industries in Japan's total imports has increased remarkably while the industry distribution of Japan's total exports has not changed much during the period. In the case of Korea, the share of high-tech and medium-tech exports has increased while the share

[1] As described in Section 3.2.2 below, Japan's outbound FDI figures are calculated based on the amount of sales of the overseas affiliates of Japanese MNEs taken from RIETI's *JIP Database 2006 Provisional Version* (RIETI, 2006). See Section 3.2.2 for details.

Table 2B. Share of each country in Japanese outbound FDI, exports, and imports: By technology level

Technology Level	*Shares in Total Japanese Outbound FDI by Technology Level (%)*											
	To CHINA			To KOREA			To USA			To World		
	1990	2002	Change	1990	2002	Change	1990	2002	Change	1990	2002	Change
High	1.04	6.36	5.33	3.25	1.16	-2.09	39.74	36.10	-3.63	25.89	30.66	4.77
Upper Middle	0.19	5.57	5.38	5.85	2.18	-3.67	42.95	37.71	-5.24	50.80	51.47	0.67
Lower Middle	0.22	6.98	6.76	7.07	6.02	-1.05	39.36	41.02	1.66	16.36	11.44	-4.93
Low	2.09	12.04	9.95	8.90	4.61	-4.29	32.28	27.56	-4.72	6.95	6.44	-0.51
Mfg. Total	0.55	6.43	5.88	5.59	2.71	-2.88	40.79	41.28	0.49	100.00	100.00	0.00

Technology Level	*Shares in Total Japanese Exports by Technology Level (%)*											
	To CHINA			To KOREA			To USA			To World		
	1990	2002	Change	1990	2002	Change	1990	2002	Change	1990	2002	Change
High	1.02	10.20	9.18	5.22	8.35	3.12	36.91	23.53	-13.38	22.17	21.86	-0.30
Upper Middle	1.82	7.67	5.85	5.52	5.11	-0.41	31.97	33.26	1.29	61.35	62.93	1.58
Lower Middle	4.52	14.13	9.60	9.71	12.98	3.26	24.23	16.49	-7.74	12.85	12.49	-0.36
Low	5.71	28.02	22.31	6.96	7.13	0.17	15.68	13.95	-1.72	3.63	2.72	-0.91
Mfg. Total	2.13	9.58	7.45	6.05	6.85	0.81	31.48	28.51	-2.97	100.00	100.00	0.00

Technology Level	*Shares in Total Japanese Imports by Technology Level (%)*											
	From CHINA			From KOREA			From USA			From World		
	1990	2002	Change	1990	2002	Change	1990	2002	Change	1990	2002	Change
High	1.12	18.12	17.01	5.29	8.03	2.75	50.87	21.42	-29.45	11.20	23.87	12.66
Upper Middle	2.01	15.46	13.45	4.43	4.73	0.31	35.49	28.49	-7.00	25.23	27.60	2.37
Lower Middle	3.97	23.88	19.92	8.13	8.75	0.62	14.85	11.35	-3.50	32.27	20.83	-11.45
Low	10.34	34.01	23.67	8.61	2.69	-5.92	22.30	15.10	-7.20	31.29	27.71	-3.59
Mfg. Total	5.15	22.99	17.84	7.03	5.79	-1.24	26.43	19.52	-6.90	100.00	100.00	0.00

Note: Highlighted cells indicate the top shares.

Sources: Ministry of Finance, Trade Statistics; RIETI (2006), JIP Database 2006 Provisional Version.

distribution in total imports has been relatively stable. These different patterns between Korea and Japan may suggest that Japanese firms are more actively involved in outsourcing to China than Korean firms, which may be partly due to differences in domestic market size and the length of experience in China. Alternatively, one may argue that Korean affiliates in China are more local-market oriented while Japanese affiliates in the country are more likely to be responsible only for one or more stages of production in parent firms' international division of labor.[1]

3. The Economic Impact of Outbound FDI and Trade: Evidence from Micro-data

3.1. Outbound FDI, Trade and Domestic Activities

The recent surge in outward FDI has sparked a debate on whether home economies are hollowing out as productive capacity is moving abroad, thus hurting domestic growth and employment. The relationship between FDI and domestic activities, especially employment, productivity and domestic investment, has been the subject of a growing body of literature,[2] which so far has produced mixed results. Hence, the relationship between FDI and domestic activities remains an unresolved issue.

Using firm-level data from U.S. and Swedish multinationals, Blomström, Fors, and Lipsey (1997), for example, examined the effect of affiliates' net sales on the employment of the parent company. They found that U.S. parent firms tend to substitute foreign production in developing countries for home employment, but this is not the case for foreign production in developed countries. On the other hand, in the case of Swedish firms, the authors provide evidence that production by foreign affiliates raises the demand for labor at home, regardless of FDI locations. More recent studies, such as Muendler and Becker (2006) and Harrison and McMillan (2006), which were already mentioned above, also provide evidence which is consistent with the results obtained by Blomström, Fors and Lipsey (1997) for the United States. In a different context, Feldstein (1995) provides aggregate level evidence that foreign investment diverts resources from domestic investment in OECD economies. On the other hand, Desai, Foley and Hines (2005a), using data on U.S. multinationals, found that foreign and domestic investments were positively correlated.[3]

As for the impact of outsourcing or overseas production on productivity at home, there are, as mentioned earlier, several studies which provide evidence of a positive relationship.

[1] For details on the degree of localization of Japanese affiliates in China, see Ito and Fukao (2007).

[2] Ahn *et al.* (2005) provide a literature survey of other strands of studies on FDI, such as on the relationship between outbound FDI and trade.

[3] Recently, Desai, Foley, and Hines (2005b) highlighted the fact that foreign and domestic operations are jointly determined by other economic factors. This indicates that endogeneity is likely to be a significant problem in regressions of foreign operations on domestic activities, unless these other determinants are explicitly controlled for in the estimation. Desai, Foley, and Hines (2005b) address this issue by employing an instrumental variable approach in analyzing the detailed affiliate-level data of U.S. multinationals. They find that the estimated effect of foreign investment on domestic investment has a larger positive magnitude in the instrumental variables equation than the corresponding estimated coefficient in the OLS equation. This result implies that omitted variables have the effect of making foreign and domestic investment look more like substitutes than is really the case.

Examples of such studies include Amiti and Wei (2006) and Görg, Hanley, and Strobl (2005). Barba Navaretti and Castellani (2004), using data on Italian multinationals, also find that FDI improves TFP and output growth, but find no significant effects on employment in the investing firm. Similarly, a recent study by Barba Navaretti, Castellani, and Disdier (2006) examining how outbound FDI to low-wage countries affects the home activities of a sample of French and Italian firms, find a positive effect on the efficiency of home activities in the case of Italy. As for Japan, Kimura and Kiyota (2006) provide evidence that firms that engage in outbound FDI exhibit higher productivity growth, although Hijzen, Inui, and Todo (2006) find no significant positive effect of FDI on productivity.

All these studies examine the substitutability or complementarity of international production and domestic activities within a firm, and many of them provide some evidence that international production and domestic activities are complementary. However, this does not alleviate potential concerns that outbound FDI or outsourcing may have a negative impact on other firms or that increasing foreign activities and outsourcing at the industry level may reduce efficiency through decreasing region- and industry-level scale or agglomeration effects. Bernard, Jensen, and Schott (2006) show that there is a negative correlation between exposure to imports from low-wage countries and plant survival and employment growth.[7] Moreover, they find a positive correlation between exposure to imports from low-wage countries and plant industry switching to industries with less exposure to imports from low-wage countries and with greater capital- and skill-intensity. If such phenomena are also occurring in Korea and Japan, we would expect a negative correlation between the increasing importance of China in outbound FDI and trade and employment at home. The study by Ahn *et al.* (2005) suggests that outbound FDI to China has a negative impact on the employment growth and TFP growth of manufacturing establishments in Korea. And on Japan, Ito (2005) has found a negative correlation between exposure to low-wage country imports and employment growth in manufacturing firms.

Below, we conduct regression analyses in order to examine how the degree or speed of industry-level globalization as well as plant/firm-specific characteristics affect employment growth and productivity growth of plants/firms in Korea and Japan. In particular, we attempt to distinguish between the impact of transactions with China – a low-wage country – and the impact of transactions with other countries. However, first it is necessary to briefly introduce the data we use.

[7] Freeman and Katz (1991), Sachs and Shatz (1994), and Revenga (1992) also find a negative correlation between employment growth and either imports (in the former two studies) or changes in import prices (in the latter study).

3.2. Data Description

3.2.1. Data for Korea

For the empirical analysis of the economic impact of Korea's outbound FDI and trade, we have constructed a plant- and industry-level dataset for the Korean manufacturing sector covering the period from 1990 to 2003. This dataset is based on three sources of information: the *Annual Report on Mining and Manufacturing Survey* (Korean National Statistical Office), the *UN Commodity Trade Statistics Database* (United Nations Statistics Division), and the *Overseas Direct Investment Statistics Yearbook* (The Export-Import Bank of Korea).

The *Mining and Manufacturing Survey* is conducted annually by the Korea National Statistical Office. The survey covers all plants with five or more employees in the mining and manufacturing industries and contains plant-level information on output, input, and a variety of additional items, including the 5-digit Korean Standard Industry Classification (KSIC) code assigned to each plant based on its major product. Variables such as plant-level employment growth, the capital-labor ratio, the ratio of non-production- to production-workers, labor productivity, and total factor productivity were calculated at the plant-level based on the information from this *Survey*. While the *Survey* covers all plants with five or more employees, we only use a subset of the data covering all manufacturing plants with no less than 50 employees in order to make it more comparable with the Japanese data.

The *UN Commodity Trade Statistics Database* ("UN COMTRADE") is compiled by the UN Statistics Division and contains annual amounts of imports, exports, and re-exports in U.S. dollars by commodity and by trading partner. Commodities are classified according to the International Trade Classification (SITC: Rev. 1 from 1962, Rev. 2 from 1976 and Rev. 3 from 1988) and the Harmonized System (HS) (from 1988 with revisions in 1996 and 2002). Imports from and exports to Korea's major trading partners by commodity based on the SITC Rev. 3 and on the HS system from 1990 to 2003 were downloaded from the UN COMTRADE website [http://unstats.un.org/unsd/COMTRADE/].

The *Overseas Direct Investment Statistics Yearbook* is published by the Export-Import Bank of Korea, an official export credit agency providing comprehensive credit and guarantees for trade and overseas investment. The *Yearbook* reports the flows and stock of outbound foreign direct investment by industry and by destination. The Export-Import Bank has its own code for industry classification ("EXIM code") which by and large is comparable to the 3-digit KSIC code. For example, the manufacturing sector as a whole consists of 71 industries according to the 3-digit KSIC code and of 70 industries according to the EXIM code. Information on annual FDI flows and stocks disaggregated by the EXIM code and by destination was downloaded from the Bank's website [http://www.koreaexim.go.kr/en/].

While the *Manufacturing Survey* contains plant-level information, the trade and FDI databases do not provide plant-level information. Therefore, to merge these three different sources, we can link the data only at a certain level of industry-wide aggregation. As the *Manufacturing Survey* and UN COMTRADE have more detailed industry classification codes, we convert their industry classification codes and match them to the EXIM code.

3.2.2. Data for Japan

For the empirical analysis for Japan, we constructed a firm- and industry-level dataset for the Japanese manufacturing sector covering the period from 1994 to 2002. This dataset is based on two sources of information: the *Basic Survey of Japanese Business Structure and Activities* (Ministry of Economy, Trade and Industry, METI) and the *JIP Database 2006 Provisional Version* (RIETI, 2006).[8] While the dataset for Korea consists of plant-level panel data, we had to use firm-level panel data for Japan because Japanese plant-level data (from the *Census of Manufactures* published by METI) do not include information on exports, imports, research and development (R&D) expenditures, and the number of non-production workers.

METI first conducted the *Basic Survey of Japanese Business Structure and Activities* in 1992 (data for fiscal 1991) and has been conducting the survey annually since 1995 (data for fiscal 1994). We did not use data from the 1992 survey because the format of that survey is significantly different from the surveys in other years and it is difficult to connect the data. The survey covers all firms with at least 50 employees or 30 million yen of paid-in capital in the Japanese manufacturing, mining, commerce, and several other services sectors. The survey contains detailed information on firm-level business activities such as the 3-digit industry in which the firm operates, its number of employees (including a breakdown by sections within the firm), sales, purchases, exports and imports (including a breakdown of the destination of sales and exports and the origin of purchases and imports), R&D and patents, the number of domestic and overseas subsidiaries, and various financial data items such as costs, profits, investment, and assets. Variables such as employment growth, the capital-labor ratio, the ratio of non-production to production workers, and TFP were calculated at the firm-level based on information from this survey.

Information on industry-level outbound FDI, exports, and imports was taken from the *JIP Database 2006 Provisional Version* (RIETI, 2006). The *JIP Database 2006* contains sector-level information on 108 sectors from 1970-2002 that can be used for TFP analyses. These sectors cover the whole Japanese economy. The Supplementary Tables of the *JIP Database 2006* provide various data items that can be used to examine the determinants of productivity growth. Data on the trade in goods by industry and country and on outbound FDI by industry and country/region are included in the Supplementary Tables. The JIP trade statistics are compiled using the customs data provided by the Ministry of Finance. As the UN COMTRADE data are compiled basically using the customs data, we believe that the JIP trade statistics are consistent with the UN COMTRADE data. Japan's outbound FDI figures are calculated based on the amount of sales of the overseas affiliates of Japanese MNEs taken from RIETI's *JIP Database 2006 Provisional Version* (RIETI, 2006). The Ministry of Finance (Ministry of Finance, 2006) provides data on *gross* investment amounts at the two-digit industry level. However, investment data at a more detailed industry level are not available. As we need outbound FDI data at a more detailed industry level for the analysis in the latter half of this paper, we used the data taken from RIETI's JIP database. Although the RIETI database does not include the amount of outstanding investment abroad, it does include the sales and employment of Japanese overseas affiliates by industry and by region. The sales by overseas affiliates of Japanese firms can be considered as the result of investment and activities by the affiliates in the past, which in this sense is a concept close to that of outward FDI stock. Moreover, the regional distribution of cumulative investment

[8] The compilation of the micro-data of the METI survey and the JIP Database 2006 was conducted as a part of the RIETI research project "Study on Industry-Level and Firm-Level Productivity in Japan" for fiscal 2004-05. The provisional version of the JIP Database 2006 is available from the RIETI website.

outflows is quite similar to that of the sales amount of Japanese affiliates. Therefore, in the case of Japan, we use sales data as a proxy for the size of Japan's outward FDI stock.

While the *Basic Survey of Japanese Business Structure and Activities* contains firm-level information, the trade and FDI database (the *JIP Database 2006*) does not. In order to link these different sources, we match the data using the JIP industry code. The industry code of the *Basic Survey* (with approximately 60 manufacturing industries) roughly corresponds to the 3-digit industry code, while the JIP industry classification includes 52 manufacturing industries.

In the sections below, we conduct regression analyses on the impact of outbound FDI and trade on employment growth and TFP growth. To summarize the data description, the analysis of Korea is based on plant- and industry-level data for the period from 1990 to 2003 for 70 manufacturing industries, while the analysis of Japan is based on firm- and industry-level data for the period from 1994 to 2002 for 52 manufacturing industries. Using these datasets, we run regressions for plant-level (firm-level) employment growth and productivity growth on plant-level (firm-level) variables as well as on industry-level variables including FDI and trade-related variables.

3.3. The Impact of FDI and Trade on Employment Growth

We begin by reporting the results of various regressions on the determinants of plant-level employment growth. Given the close link between FDI and trade described earlier, both outbound FDI and trade growth (or, alternatively, export and import growth) are included as explanatory variables. As we have seen in Section 2, the major destinations for outbound FDI from Korea are China and the U.S., while the major trading partners are China, Japan, and the U.S. As for Japan, the major destinations for outbound FDI are the U.S. and China, while the major trading partners are the U.S., China, and Korea. Although Korea's share in Japanese trade is not as large as the shares of China and the U.S., Korea is still one of Japan's major trading partners. Moreover, as shown in Tables 2A and 2B in Section 2, the composition of high-tech, medium-tech, and low-tech industries varies substantially from country to country and also over time. To take such variations into account, we include the shares of the major partner countries in FDI and trade as additional explanatory variables. The growth rates and country shares of FDI and trade are measured at a relatively detailed industry level based on the EXIM code with 70 manufacturing industries in the case of Korea and on the JIP code with 52 manufacturing industries in the case of Japan.

Our regression equation for employment growth is:

$$\frac{\ln L_{i,t+3} - \ln L_{i,t}}{3} = \alpha_0 + \alpha_{Plant} \cdot X_{i,t} + \alpha_{Industry} \cdot Y_{j,t} + \alpha_{D_t} \cdot D_t + u_i + \varepsilon_{i,t} \qquad (1)$$

where the left-hand-side variable is the average annual employment growth rate in terms of the number of workers at plant (firm) i from year t to year (t +3) and the right-hand-side variables are as follows:[9]

[9] A description of the variables is provided in Appendix Table 1.

- $X_{i,t}$: a vector of plant- or firm-specific variables for plant (firm) i in year t, which includes the growth rate of shipments (sales), the share of non-production workers, the capital-labor ratio, R&D intensity measured as R&D expenditure divided by shipments or sales, the export-sales ratio, and the number of workers.
- $Y_{j,t}$: a vector of industry-specific variables for industry j to which plant (firm) i belongs in year t, including the industry-level non-production worker share, the capital-labor ratio, R&D intensity, export intensity, import intensity, and the growth rates of outward FDI, exports, imports, and trade (exports plus imports). Moreover, in order to examine the impact of FDI to or trade with major partners, we include the industry-level share of each destination or partner: the shares of FDI to China, the U.S., Japan, and Korea, the shares of exports to China, the U.S., Japan, and Korea, and the shares of imports from China, the U.S., Japan, and Korea.
- D_t: a vector of year dummy variables.
- u_i: plant (firm) specific fixed effects.

We estimate Equation (1) for both Korea and Japan applying a fixed-effect panel regression.[10] In the case of Korea, we focus on plants with 50 or more workers in order to increase the comparability of results with those for Japan.[11] Table 3 summarizes the fixed-effect panel estimation results for both Korea and Japan. Columns 1-3 are the estimation results for Korea, and columns 4-6 are the estimation results for Japan.

According to the results reported in Table 3, it appears that sales growth and the export-sales ratio at the plant level (firm level) have significant positive effects on employment growth at the plant level (firm level) both in Korea and Japan. On the other hand, for both countries, the logarithm of the initial size of the plant (firm) measured by the number of workers has a negative effect on employment growth, reflecting the trivial fact that smaller plants (firms) have more room for further growth than larger plants (firms). As for other plant-level (firm-level) explanatory variables, the share of non-production workers and R&D intensity have significant negative effects on employment growth in Korea, but no significant effects on employment growth in Japan.

We are now ready to focus on the main variables of interest. Columns 1 and 4 show the estimation results for the specification including the industry-level FDI growth rate and the industry-level share of FDI for each destination. In the case of both Korea and Japan, the growth rate of outbound FDI at the industry-level does not have a significant coefficient. On the other hand, the estimated coefficient for the share of FDI to China is positive and statistically significant in the Korea estimation (column 1), while it is negative and statistically significant in the Japan estimation (column 4). As described in Section 2, China is

[10] The Hausman test results support the fixed-effect model rather than the random-effect model.

[11] However, we should still note that the analysis for Korea is based on plant-level data, while the analysis for Japan is based on firm-level data. In fact, most firms (more than 90% in the case of Korea) own only one plant, in which case plant-level data and firm-level data should be almost identical. However, larger firms are likely to have more than one plant.

Table 3. Fixed-effect regression results for the 3-year average employment growth rate (overlapped)

	Korea			Japan		
	(1)	(2)	(3)	(4)	(5)	(6)
(Sales growth, 3-year)$_{f,t}$	**0.10758*** **	**0.10749*** **	**0.10711*** **	**0.17023*** **	**0.17426*** **	**0.17391*** **
	(16.54)	(16.29)	(15.95)	(21.47)	(23.24)	(23.23)
(Non-production worker share)$_{f,t}$	**-0.00366** **	**-0.00359** **	**-0.00361** **	**-0.00099**	**0.00094**	**0.00092**
	(-2.35)	(-2.30)	(-2.32)	(-0.41)	(0.44)	(0.42)
(Capital-labor ratio)$_{f,t}$	**-0.000001**	**0.000002**	**0.000001**	**0.00007**	**-0.00001**	**-0.00002**
	(-0.08)	(0.14)	(0.04)	(0.74)	(-0.10)	(-0.26)
(R&D intensity)$_{f,t}$	**-0.03132** **	**-0.03753*** **	**-0.03687*** **	**-0.02323**	**-0.02260**	**-0.02372**
	(-2.17)	(-2.71)	(-2.66)	(-0.88)	(-0.86)	(-0.91)
(Export-sales ratio)$_{f,t}$	**0.00749** *	**0.00709** *	**0.00707** *	**0.01954*** **	**0.01860*** **	**0.01805*** **
	(1.91)	(1.80)	(1.79)	(3.08)	(3.01)	(2.92)
ln(Number of workers)$_{f,t}$	**-0.33569** ``	**-0.33403*** **	**-0.33446*** **	**-0.26185*** **	**-0.26780*** **	**-0.26756*** **
	(-80.45)	(-79.73)	(-78.81)	(-52.04)	(-63.60)	(-63.56)
(Non-production worker share)$_{j,t}$	**-0,01977** *	**-0.03415*** **	**-0.02935*** **	**-0.71647*** **	**-0.68424*** **	**-0.72937*** **
	(-1.88)	(-3.09)	(-2.65)	(-4.50)	(-5.67)	(-6.08)
(Capital-labor ratio)$_{j,t}$	**-0.00006*** **	**-0.00009*** **	**-0.00007** **	**0.00003**	**-0.00014*** **	**-0.00011*** **
	(-2.68)	(-3.32)	(-2.32)	(0.72)	(-3.69)	(-2.81)
(R&D intensity)$_{j,t}$	**-0.00265**	**-0.06655**	**-0.04599**	**0.29403*** **	**0.27229*** **	**0.22197** **
	(-0.03)	(-0.76)	(-0.52)	(2.92)	(2.73)	(2.25)
(Export intensity)$_{j,t}$	**-0.01067**	**-0.01893** *	**-0.01801**	**0.000021**	**0.000027**	**0.00003**
	(-1.03)	(-1.65)	(-1.60)	(0.93)	(1.20)	(1.14)
(Import intensity)$_{j,t}$	**0.00506**	**0.00464**	**0.00475**	**-0.00001**	**-0.000003**	**0.00001**
	(1.13)	(1.04)	(1.06)	(-0.25)	(-0.13)	(0.48)
ln(FDI)$_{j,t}$-ln(FDI)$_{j,t-1}$	**0.00096**			**0.00163**		
	(0.25)			(0.54)		
(FDI to China share)$_{j,t}$	**0.02321** *			**-0.05705** **		
	(1.89)			(-2.51)		
(FDI to USA share)$_{j,t}$	**-0.00699**			**0.00889**		
	(-0.50)			(1.04)		

Table 3. (Continued)

	Korea			Japan		
	(1)	(2)	(3)	(4)	(5)	(6)
$(\text{FDI to Japan share})_{j,t}$	**0.08378**					
	(1.44)					
$(\text{FDI to Korea share})_{j,t}$				**0.03982** *		
				(1,80)		
$\ln(\text{Exports})_{j,t}-\ln(\text{Exports})_{j,t-1}$		**-0.00566**			**0.00188**	
		(-0.48)			(0.82)	
$(\text{Exp to China share})_{j,t}$		**0.08514** ***			**-0.01998**	
		(3.31)			(-1.08)	
$(\text{Exp to USA share})_{j,t}$		**0.08344** **			**0.02146**	
		(2.37)			(1.56)	
$(\text{Exp to Japan share})_{j,t}$		**-0.05210**				
		(-1.31)				
$(\text{Exp to Korea share})_{j,t}$					**0.05985** ***	
					(3.49)	
$\ln(\text{Imports})_{j,t}-\ln(\text{Imports})_{j,t-1}$		**0.00914**			**0.00271**	
		(0.55)			(1.27)	
$(\text{Imp from China share})_{j,t}$		**-0.04642**			**-0.02312**	
		(-1.59)			(-1.54)	
$(\text{Imp from USA share})_{j,t}$		**0.01039**			**-0.01728**	
		(0.26)			(-1.20)	
$(\text{Imp from Japan share})_{j,t}$		**-0.01925**				
		(-1.00)				
$(\text{Imp from Korea share})_{j,t}$					**0.02350**	
					(1.53)	
$\ln(\text{Trade})_{j,t}-\ln(\text{Trade})_{j,t-1}$			**0.005011**			**0.00285**
			(0.26)			(1,03)
$(\text{Trade with China share})_{j,t}$			**0.02022**			**-0.06672** ***
			(0.72)			(-3.19)

Table 3. (Continued)

	Korea						Japan					
	(1)		(2)		(3)		(4)		(5)		(6)	
(Trade with USA share)$_{j,t}$					**0.04594**						**-0.04871**	***
					(1.31)						(-2.83)	
(Trade with Japan share)$_{j,t}$					**-0.00058**							
					(-0.03)							
(Trade with Korea share)$_{j,t}$											**0.00447**	
											(0.20)	
Intercept	**1.61520**	***	**1.58640**	***	**1.58310**	***	**1.63900**	***	**1.66738**	***	**1.70152**	***
	(76.60)		(69.33)		(69.08)		(26.68)		(35.09)		(35.87)	
Number of observations	37,930		37,160		37,160		29,077		37,354		37,354	
F value	563.1	***	458.7	***	528.7	***	256.0	***	288.8	***	343.2	***
R-squared	0.7386		0.7352		0.7347		0.4351		0.4455		0.4450	

Note: Heteroskedasticity-robust t-statistics in parentheses.

the most important destination of outbound FDI in medium-tech and low-tech industries for both Korea and Japan. So far, the estimation results suggest that plant-level (firm-level) employment growth in these industries is higher than the average in Korea but lower than the average in Japan after controlling for plant-specific (firm-specific) fixed effects.

Columns 2 and 5 show the estimation results of the specification including industry-level trade related variables (export growth, export shares, import growth, and import shares). The industry-level import and export growth rates do not have any significant impact on employment growth at the plant-level (firm-level) in both countries. As for import competition from a low-wage country, i.e., China, there is no significant negative effect both in Korea and Japan (columns 2 and 5) although the sign of the coefficient is negative for both countries. On the other hand, the shares of exports to China and the U.S. have a significant positive impact on plant-level employment growth in Korea, while the share of exports to Korea has a significant positive impact on firm-level employment growth in Japan.

Finally, columns 3 and 6 show the estimation results when trade growth (the sum of imports and exports) and trade shares are used as industry-level explanatory variables. The results indicate that industry-level trade growth does not have a significant impact on plant-level (firm-level) employment growth, which is consistent with the estimation results in columns 2 and 4. The shares of trade with China and the U.S. appear to have a significantly negative effect in Japan but do not affect plant-level employment growth in Korea. These results imply that industry-level globalization does not have any significant negative impact on plant-level employment growth in Korea if plant-specific fixed effects are controlled for. However, in the case of Japan, firm-level employment growth is significantly lower on average in industries where the shares of FDI to and trade with China are larger, even when controlling for firm-specific fixed effects. However, as the share of trade with the U.S. also has a significant negative impact on firm-level employment growth in Japan, we cannot conclude that the decrease in firm-level employment in Japan is attributable to competition from China.

3.4. Impacts of FDI and Trade on Productivity Growth

Turning from job-creation to productivity growth, we investigate the impact of FDI and trade on productivity growth using a similar regression approach as above. Our regression equation for total factor productivity (*TFP*) growth is:

$$\frac{\ln TFP_{i,t+3} - \ln TFP_{i,t}}{3} = \beta_0 + \beta_{Plant} \cdot X_{i,t} + \beta_{Industry} \cdot Y_{j,t} + \beta_{D_t} \cdot D_t + u_i + \varepsilon_{i,t} \qquad (2)$$

where the left-hand-side variable is the average annual log growth rate of total factor productivity (*TFP*) at plant (firm) i from year t to year (t +3) and the right-hand-side variables are the same as those in Equation (1) except that $X_{i,t}$ here contains the level of ln *TFP* for plant (firm) i at year t instead of the sales growth rate.

Following Good, Nadiri, and Sickles (1999), Aw, Chen, and Roberts (2001), Hahn (2005), and Ahn, Fukao, and Kwon (2004), plant-level total factor productivity (TFP) is estimated by the chained-multilateral index number approach. This uses a separate reference point for each cross-section of observations and then chain-links the reference points together over time as in the Tornqvist-Theil index. The output, input, and productivity level of each plant in each year is measured relative to the hypothetical plant at the base-time period. This approach allows us to make transitive comparisons of productivity levels among observations in a panel dataset. The productivity index for plant (firm) *i* at time *t* is measured as follows:

$$\ln TFP_{it} = (\ln Y_{it} - \overline{\ln Y_t}) + \sum_{\tau=2}^{t} (\overline{\ln Y_\tau} - \overline{\ln Y_{\tau-1}})$$

$$- \{ \sum_{n=1}^{N} \frac{1}{2}(S_{nit} + \overline{S_{nt}})(\ln X_{nit} - \overline{\ln X_{nt}}) + \sum_{\tau=2}^{t} \sum_{n=1}^{N} \frac{1}{2}(\overline{S_{n\tau}} + \overline{S_{n\tau-1}})(\overline{\ln X_{n\tau}} - \overline{\ln X_{n\tau-1}}) \} \quad (3)$$

where *Y*, *X*, *S*, and *TFP* denote output, input, the input share, and the TFP level, respectively, and symbols with an upper bar are the corresponding measures for the hypothetical firms. The subscripts τ and *n* are indices for time and inputs, respectively.

We estimate Equation (2) for both Korea and Japan applying a fixed-effect panel regression.[38] Table 4 summarizes the fixed-effect panel estimation results for both Korea and Japan. Columns 1-3 show the estimation results for Korea, while columns 4-6 show those for Japan.

The natural log of the *TFP* level at year *t* is the so-called convergence term in a typical growth regression. In our regressions, the estimated coefficient was consistently negative. The logarithm of the initial size of the plant measured by the number of workers has a positive effect on *TFP* growth in the case of Korea, reflecting economies of scale. In contrast, this variable does not have a significant coefficient in the case of Japan. Other plant-level (firm-level) variables do not have any significant coefficients both for Korea and Japan when plant-specific (firm-specific) fixed effects are controlled for. At the industry-level, we have controlled for the share of non-production workers, the capital-labor ratio, R&D intensity, and export and import intensities. In the case of Korea, the industry-level *export* intensity has a negative impact on TFP growth, while in the case of Japan, it is the industry-level *import* intensity that has a negative effect on TFP growth. In Korea, the average productivity growth rate is low in industries with a high export intensity. One possible explanation is that many plants face fierce export competition; alternatively, the negative effect may be partly related to the depreciation of the won in the wake of the 1997/98 economic crisis, which may have raised exporters' price competitiveness in the global market without forcing them to make efforts to improve efficiency. In the case of Japan, the negative effect of import intensity may mean that in industries where import intensity is high, the productivity growth rate is lower

[38] Again, the Hausman test results support the fixed-effect model rather than the random-effect model.

[40] We did not include the firm-level import-sales ratio as an explanatory variable because plant-level import-sales ratios are not available for Korea. Estimation results including the firm-level import-sales ratio for Japan are mostly consistent with the results reported in Tables 3 and 4. The firm-level import-sales ratio has a significant positive coefficient in the employment growth equations, while it does not have a significant coefficient in the TFP growth equations.

Table 4. Fixed-effect regression results for the 3-year average TFP growth rate (overlapped)

	Korea			Japan		
	(1)	(2)	(3)	(4)	(5)	(6)
$\ln(TFP)_{f,t}$	**-0.35486** ***	**-0.35495** ***	**-0.35493** ***	**-0.35088** ***	**-0.35030** ***	**-0.35339** ***
	(-108.48)	(-106.53)	(-106.64)	(-87.36)	(-94.82)	(-91.26)
(Non-production worker share)$_{f,t}$	**0.00006**	**0.00011**	**0.00010**	**0.00101**	**0.00066**	**0.00048**
	(0,08)	(0.15)	(0.13)	(0.75)	(0.56)	(0,41)
(Capital-labor ratio)$_{f,t}$	**-0.0000002**	**0.0000003**	**0.0000001**	**-0.00004**	**0.00005**	**0.00001**
	(-0.02)	(0.03)	(0.01)	(-0.83)	(1.09)	(0,12)
(R&D intensity)$_{f,t}$	**0.01002**	**0.00886**	**0.00929**	**-0.01557**	**-0.01648**	**-0.01489**
	(0,93)	(0.82)	(0.85)	(-0.92)	(-1.01)	(-0.91)
(Export-sales ratio)$_{f,t}$	**0.00184**	**0.00129**	**0.00146**	**-0.00203**	**-0.00234**	**-0.00192**
	(0,54)	(0.37)	(0.42)	(-0.55)	(-0.65)	(-0.53)
ln(Number of workers)$_{f,t}$	**0.00429** **	**0.00453** ***	**0.00439** **	**0.00194**	**0.00012**	**0.00029**
	(2,54)	(2.64)	(2.55)	(0.98)	(0.07)	(0,17)
(Non-production worker share)$_{j,t}$	**-0.00657**	**-0.00119**	**-0.00229**	**1.73361** ***	**1.49530** ***	**1.53345** ***
	(-0.85)	(-0.15)	(-0.29)	(21.58)	(25.91)	(27,74)
(Capital-labor ratio)$_{j,t}$	**-0.00002**	**-0.00002**	**-0.00002**	**0.00002**	**0.00007** ***	**-0.00003**
	(-1.17)	(-1.13)	(-1.07)	(0,91)	(4.20)	(-1.45)
(R&D intensity)$_{j,t}$	**0.13987** **	**0.14960** **	**0.13892** **	**-1.31927** ***	**-1.28090** ***	**-1.30055** ***
	(2,15)	(2.27)	(2.12)	(-22.39)	(-22.38)	(-22.57)
(Export intensity)$_{j,t}$	**-0.01552** *	**-0.01782** **	**-0.01887** **	**-0.00001**	**-0.00001**	**-0.00003** **
	(-1.87)	(-2.07)	(-2.23)	(-1.02)	(-0.99)	(-2.34)
(Import intensity)$_{j,t}$	**0.00671**	**0.00634**	**0.00616**	**-0.00003** **	**-0.00004** ***	**-0.00004** ***
	(1.43)	(1.36)	(1.32)	(-2.51)	(-3.12)	(-3.48)
$\ln(FDI)_{j,t}-\ln(FDI)_{j,t-1}$	**-0.00303**			**0.00713** ***		
	(-0.85)			(3.83)		

Table 4. (Continued)

	Korea		Japan			
	(2)	(3)	(4)	(5)	(6)	(1)
(FDI to China share)$_{j,t}$	**0.00007**			**-0.04372** ***		
	(0,01)			(-3.52)		
(FDI to USA share)$_{j,t}$	**0.01751** *			**0.00587**		
	(1,75)			(1,15)		
(FDI to Japan share)$_{j,t}$	**-0.07716** *					
	(-1.33)					
(FDI to Korea share)$_{j,t}$				**0.10530** ***		
				(9,10)		
ln(Exports)$_{j,t}$-ln(Exports)$_{j,t-1}$		**-0.00649**			**-0.01199** ***	
		(-0.83)			(-9.40)	
(Exp to China share)$_{j,t}$		**0.03319** *			**0.04248** ***	
		(1,70)			(4.37)	
(Exp to USA share)$_{j,t}$		**0.05577** *			**0.03519** ***	
		(1,96)			(5.11)	
(Exp to Japan share)$_{j,t}$		**0.00989**				
		(0,47)				
(Exp to Korea share)$_{j,t}$					**0.06768** ***	
					(6.36)	
ln(Imports)$_{j,t}$-ln(Imports)$_{j,t-1}$		**-0.00996**			**-0.00834** ***	
		(-0.84)			(-8.16)	
(Imp from China share)$_{j,t}$		**-0.04605** **			**-0.08005** ***	
		(-2.26)			(-10.21)	
(Imp from USA share)$_{j,t}$		**-0.09141** **			**-0.01574** ***	
		(-2.53)			(-2.16)	
(Imp from Japan share)$_{j,t}$		**-0.02355** *				
		(-1.81)				

Table 4. (Continued)

	Korea				Japan	
	(1)	(2)	(3)	(4)	(5)	(6)
(Imp from Korea share)$_{j,t}$					**-0.07440** ***	
					(-10.03)	
ln(Trade)$_{j,t}$-ln(Trade)$_{j,t-1}$			**-0.00729**			**-0.00130**
			(-0.58)			(-0.92)
(Trade with China share)$_{j,t}$			**-0.02058**			**-0.03267** ***
			(-1.06)			(-2.98)
(Trade with USA share)$_{j,t}$			**-0.01457**			**0.06330** ***
			(-0.53)			(8.46)
(Trade with Japan share)$_{j,t}$			**-0.01559**			
			(-1.01)			
(Trade with Korea share)$_{j,t}$						**0.12520** ***
						(9.18)
Intercept	**0.08924** ***	**0.10092** ***	**0.09908** ***	**-0.53792** ***	**-0.44784** ***	**-0.48249** ***
	(9.17)	(8.77)	(8.98)	(-18.25)	(-20.41)	(-22.54)
Number of observations	30,076	29,475	29,475	29,077	37,354	37,354
F value	675.1 ***	550.9 ***	649.7 ***	468.1 ***	486.6 ***	210.2 ***
R-squared	0.5981	0.5967	0.5965	0.5984	0.5932	0.2953

Note: Heteroskedasticity-robust t-statistics in parentheses.

than the average, which may in turn indicate a negative impact from import competition or a "hollowing-out" effect.[40]

The industry-level R&D intensity affects TFP growth positively in Korea but negatively in Japan. In the case of Korea, industry-level R&D activities contribute to average plant-level productivity growth probably through R&D spillover effect, while in the case of Japan, the negative coefficient of the industry-level R&D intensity may be an indication of high competitive pressures in R&D-intensive industries. A likely explanation for the difference in the contribution of industry R&D to the productivity growth of individual plants/firms is the difference in the distance to the technological frontier. As most Korean manufacturing sectors are still operating some way off the technological frontier, technology is relatively simple and easy to imitate. Industry-level R&D intensity thus is likely to produce greater spillover effects, which in turn lead to the positive contribution to productivity growth. In contrast, most Japanese manufacturing sectors are operating very close to the technological frontier; consequently, technology is more sophisticated and not easily imitated quickly, technological competition among firms is fierce, and the competition effect dominates any spillover effects.

In the case of Japan, the growth rate of outbound FDI at the industry-level has a positive coefficient with a sufficiently high heteroscedasticity-robust *t*-ratio, suggesting that an increase in outbound FDI tends to have a positive effect on productivity growth (column 4). However, the estimated coefficient for the share of FDI to China is negative and statistically significant (with a high heteroscedasticity-robust *t*-ratio) for Japan, which suggests the industry-level expansion of FDI to China may lower the average firm-level productivity growth rate. In contrast, in the case of Korea, the growth rate of outbound FDI at the industry-level does not have a significant coefficient (column 1). The coefficient for the share of FDI to China is also not statistically significant, while the coefficient for the share of FDI to the U.S. is positive and statistically significant. Although the employment growth regression results showed that plant-level employment growth was higher on average in industries where a larger share of outbound FDI goes to China, plant-level TFP growth is not significantly affected by the share of outbound FDI in China.

The estimated coefficients for growth rates of exports and imports are all negative for both countries, although they are statistically significant only in the case of Japan (columns 2 and 5). This implies that the firm-level TFP growth rate tends to be lower on average in industries where exports and imports are growing. On the other hand, in the case of both countries, export share variables tend to have a significant positive coefficient while import share variables tend to have a significant negative coefficient regardless of the partner country (columns 2 and 5). However, the overall effects of trade growth (exports plus imports) are not statistically significant for both countries, and furthermore, the trade share variables are not statistically significant in the case of Korea (columns 3 and 6). In the case of Japan, firm-level TFP growth tends to be lower in industries with a high share of trade with China, while the shares of trade with the U.S. and Korea had the opposite impact. In the case of Japan, medium-low and low-tech industries tend to have a larger share of trade with China, while high- and medium high-tech industries tend to have a larger share of trade with the United States. Moreover, FDI to and trade with Korea have been shifting to high-tech industries. This distribution in trade shares suggests that, on average, firm-level productivity growth is negatively affected in lower-tech industries by competition from China. In the higher-technology industries, however, firm-level productivity growth on average is positively affected, probably as a result of knowledge spillover or learning effects from trade with the

United States. In addition, skill or technology upgrading in Korean high-tech industries may have contributed to firm-level productivity growth in Japan.

3.5. Discussion: The Impact of FDI and Trade on Domestic Performance

According to the regression results, both in Korea and Japan, plants (or firms) with a high export-sales ratio show higher employment growth, although they do not show any significantly higher TFP growth.[41] However, at present, we cannot conclude that at the industry-level, greater exposure to foreign markets significantly contributes to job creation and improvements in productivity either in Korean plants or Japanese firms when controlling for plant-specific (firm-specific) fixed effects. Rather, industry-level export intensity tends to have a negative impact on the TFP growth of Korean plants. Moreover, industry-level import intensity and export and import growth tend to have a negative impact on TFP growth at Japanese firms. Although there is no strong negative effect of competition from China in the case of Korea, FDI to and trade with China have a significantly negative impact on employment and TFP growth in the case of Japan. These results thus provide some evidence of a negative impact of competition from China, in the case of Japan.

On the other hand, in the case of Korea, the fixed-effect panel estimation results in this paper do not provide any strong evidence of a negative impact of competition from China. This result is at odds with the findings of Ahn *et al.* (2005), who showed that the higher the share of outward FDI to China in a certain industry, the lower the average employment growth and TFP growth of establishments in that industry. The different results may come from the difference in estimation methodologies, i.e., the use of fixed-effect panel regression in this study and of ordinary least square (OLS) regression in Ahn *et al.* (2005). Tables 5 and 6 show the estimation results for Equations (1) and (2) above, pooling all observations and applying OLS for both Korea and Japan. Table 5 presents the results for employment growth, while Table 6 shows those for TFP growth. This time, we include broad industry dummy variables (16 manufacturing industries) in order to capture unobserved industry-specific factors instead of plant-specific (firm-specific) fixed effects both in the employment growth and the TFP growth equation. Looking at the country share variables in Tables 5 and 6 in the results for Korea, we find that the industry-level share of FDI to China and the share of trade with China have a significant negative impact both on plant-level employment and on TFP

[41] Many empirical studies have investigated the performance (including productivity) of exporters to examine whether their performance is better than that of non-exporters and to determine where performance differences come from, i.e., from self-selection effects, learning-by-exporting effects, etc. The self-selection effect refers to the effect by which more productive plants self-select into the export market. The learning-by-exporting effect, on the other hand, refers to the effect by which exporters learn from knowledgeable buyers who provide them with blueprints and then offer technical assistance. A substantial number of studies have shown that in many countries exporters tend to have high productivity. However, in the case of Korea, the empirical evidence on these issues is rather mixed. Aw, Chung, and Roberts (2000), for example, did not find any strong evidence of self-selection or learning-by-exporting in Korea. On the other hand, Hahn (2005) did find that exporters enjoy higher productivity and employment growth than non-exporters, although the results were somewhat sensitive to which time period was chosen for the analysis. Ahn, Fukao, and Kwon (2004) in their study on Korea for the period from 1990 to 1998 found that plants with a higher export-ratio showed higher TFP growth. However, as Hahn (2005) emphasized, the different conclusions derived from these studies, including the present study, may well be due to differences in the time periods covered, datasets used, and methodologies employed. The issue of the performance of exporters deserves further investigation.

growth, which is consistent with the results of Ahn *et al.* (2005). On the other hand, the estimated results for Japan in Tables 5 and 6 are mostly in line with those in Tables 3 and 4.

As Ahn *et al.* (2005) showed, in the case of Korea, plant-level average employment growth and TFP growth tend to be lower in industries with a higher share of outward FDI to China. Moreover, in industries with a higher share of trade with China, plant-level average employment growth is significantly lower. However, these negative effects of transactions with China disappear if plant-specific fixed effects are controlled for. Therefore, we may conclude that in the case of Korea, effects from cross-industry variations in the share of each FDI destination or trade partner dominate effects from time-series variations in the shares. Plants in an industry with a larger exposure to competition from China in terms of trade and global outsourcing tend to show lower growth of employment and productivity when compared with plants in other industries with less exposure to competition from China. However, we do not have evidence that an increase in China's share in terms of outward FDI or trade within a certain industry has a significant negative impact on plant-level employment or TFP growth. In Japan's case, on the other hand, the share of FDI to China and share of trade with China still have a significant negative impact, particularly on the firm-level TFP growth rate, even after controlling for firm-specific fixed effects.

4. Conclusion

This paper examined the economic impact of the increase in outbound FDI to and trade with China, comparing the cases of Korea and Japan. In the comparison, we paid attention to the following aspects. First, while on the one hand, Japan has longer experience with outbound FDI and overseas production, it is Korea, whose investment in recent years has been more heavily concentrated in China. Second, Japanese firms may be more actively involved in outsourcing to China in the sense that their affiliates in China are more likely to be responsible for one or more stages of production as a part of the parent firm's international division of labor. As a result, trade between parent firms in Japan and their affiliates in China has been increasing rapidly. Third, although Korean firms are now quite competitive in the global market in the fields of electrical machinery, electronic components, communication equipment, automobiles, etc., there still exists a significant gap between Korea and Japan in productivity levels, especially in major exporting industries.

Table 5. Regression results for the 3-year average employment growth rate (overlapped)

	Korea			Japan		
	(1)	(2)	(3)	(4)	(5)	(6)
$(\text{Sales growth, 3-year})_{f,t}$	**0.26774** ***	**0.26594** ***	**0.26578** ***	**0.35763** ***	**0.36616** ***	**0.36596** ***
	(28.33)	(27.43)	(27.42)	(46.79)	(52.15)	(52.15)
$(\text{Non-production worker share})_{f,t}$	**-0.00030**	**-0.00028**	**-0.00029**	**0.00712** ***	**0.00664** ***	**0.00674** ***
	(-0.49)	(-0.45)	(-0.47)	(4.00)	(4.16)	(4.22)
$(\text{Capital-labor ratio})_{f,t}$	**0.00012** ***	**0.00012** ***	**0.00012** ***	**0.00013** ***	**0.00011** ***	**0.00011** ***
	(9.83)	(9.79)	(9.74)	(5.01)	(5.04)	(4.95)
$(\text{R\&D intensity})_{f,t}$	**-0.01652**	**-0.01963**	**-0.01978**	**0.02106**	**0.02095**	**0.02231**
	(-0.64)	(-0.76)	(-0.77)	(1.10)	(1.16)	(1.23)
$(\text{Export-sales ratio})_{f,t}$	**0.01833** ***	**0.01716** ***	**0.01681** ***	**-0.00014**	**-0.00187**	**-0.00152**
	(6.55)	(6.07)	(5.99)	(-0.04)	(-0.54)	(-0.44)
$\ln(\text{Number of workers})_{f,t}$	**-0.09227** ***	**-0.09118** ***	**-0.09122** ***	**-0.01032** ***	**-0.01025** ***	**-0.01026** ***
	(-53.30)	(-52.66)	(-52.53)	(-25.53)	(-28.23)	(-28.27)
$(\text{Non-production worker share})_{j,t}$	**0.01083** **	**0.01523** ***	**0.01864** ***	**0.00292**	**0.00189**	**-0.00116**
	(1.99)	(2.96)	(3.74)	(0.24)	(0.25)	(-0.15)
$(\text{Capital-labor ratio})_{j,t}$	**-0.00009** ***	**-0.00009** ***	**-0.00007** ***	**0.00000**	**-0.00003** **	**0.00001**
	(-5.81)	(-4.77)	(-4.48)	(-0.56)	(-2.21)	(0.47)
$(\text{R\&D intensity})_{j,t}$	**0.14872**	**0.12775**	**0.13221**	**0.00758**	**0.00327**	**-0.01166**
	(1.44)	(1.20)	(1.25)	(0.13)	(0.08)	(-0.30)
$(\text{Export intensity})_{j,t}$	**0.00335**	**0.01357**	**0.01207**	**-0.000003**	**-0.000003**	**-0.00001**
	(0.42)	(1.53)	(1.44)	(-0.38)	(-0.38)	(-0.82)
$(\text{Import intensity})_{j,t}$	**-0.00286**	**0.00164**	**-0.00140**	**0.00002** ***	**0.00002** ***	**0.00001** ***
	(-0.78)	(0.41)	(-0.38)	(3.42)	(3.79)	(2.73)
$\ln(\text{FDI})_{j,t}-\ln(\text{FDI})_{j,t-1}$	**0.00624** ***			**0.00045**		
	(2.83)			(0.13)		
$(\text{FDI to China share})_{j,t}$	**-0.01091** *			**-0.03581**		
	(-1.84)			(-1.57)		

Table 5. (Continued)

	Korea			Japan		
	(1)	(2)	(3)	(4)	(5)	(6)
$(\text{FDI to USA share})_{j,t}$	**0.02600** ***			**0.00255**		
	(3.93)			(0.55)		
$(\text{FDI to Japan share})_{j,t}$	**0.02559**					
	(1.21)					
$(\text{FDI to Korea share})_{j,t}$				**0.01175**		
				(0.90)		
$\ln(\text{Exports})_{j,t}-\ln(\text{Exports})_{j,t-1}$		**-0.01213** **			**0.00554** *	
		(-2.31)			(1.81)	
$(\text{Exp to China share})_{j,t}$		**-0.00812**			**-0.02676** *	
		(-0.66)			(-1.81)	
$(\text{Exp to USA share})_{j,t}$		**0.05778** ***			**0.01820** **	
		(4.87)			(2.51)	
$(\text{Exp to Japan share})_{j,t}$		**-0.01265**				
		(-0.99)				
$(\text{Exp to Korea share})_{j,t}$					**0.01062**	
					(1.35)	
$\ln(\text{Imports})_{j,t}-\ln(\text{Imports})_{j,t-1}$		**0.03206** ***			**0.00617** **	
		(3.71)			(2.30)	
$(\text{Imp from China share})_{j,t}$		**-0.06383** ***			**0.00108**	
		(-4.24)			(0.20)	
$(\text{Imp from USA share})_{j,t}$		**-0.01853**			**-0.00109**	
		(-1.00)			(-0.19)	
$(\text{Imp from Japan share})_{j,t}$		**-0.01451**				
		(-1.38)				
$(\text{Imp from Korea share})_{j,t}$					**0.01404** **	
					(2.30)	

Table 5. (Continued)

	Korea						Japan					
	(1)		(2)		(3)		(4)		(5)		(6)	
$\ln(\text{Trade})_{j,t}$-$\ln(\text{Trade})_{j,t-1}$					**-0.009547**						**0.00830**	**
					(-0.93)						(2.33)	
(Trade with China share)$_{j,t}$					**-0.04419**	***					**-0.00089**	
					(-3.24)						-0.09	
(Trade with USA share)$_{j,t}$					**0.02631**	*					**0.01761**	**
					(1.95)						(2.41)	
(Trade with Japan share)$_{j,t}$					**0.00567**							
					(0.48)							
(Trade with Korea share)$_{j,t}$											**-0.01053**	
											(-0.78)	
Intercept	**0.37687**	***	**0.38338**	***	**0.37613**	***	**0.05154**	***	**0.05092**	***	**0.05163**	***
	(9.75)		(9.70)		(9.53)		(9.38)		(10.47)		(11.29)	
Number of observations	37,930		37,160		37,160		29,077		37,354		37,354	
F value		***		***		***	104.9	***	111.5	***	121.6	***
R-squared	0.5776		0.5713		0.5708		0.2572		0.2499		0.2495	

Note: Heteroskedasticity-robust t-statistics in parentheses.

Table 6. Regression results for the 3-year average TFP growth rates (overlapped)

	Korea			Japan		
	(1)	(2)	(3)	(4)	(5)	(6)
$\ln(TFP)_{f,t}$	**-0.21394** ***	**-0.21508** ***	**-0.21394** ***	**-0.15079** ***	**-0.13669** ***	**-0.13477** ***
	(-81.23)	(-80.96)	(-80.25)	(-18.72)	(-18.38)	(-18.40)
(Non-production worker share)$_{f,t}$	**0.00271** ***	**0.00282** ***	**0.00278** ***	**0.00562** ***	**0.00422** ***	**0.00400** ***
	(4.50)	(4.55)	(4.50)	(5.73)	(5.17)	(4.89)
(Capital-labor ratio)$_{f,t}$	**-0.00001** **	**-0.00002** ***	**-0.00002** ***	**-0.00010** **	**-0.00004**	**-0.00003**
	(-2.55)	(-2.94)	(-2.96)	(-1.99)	(-0.91)	(-0.66)
(R&D intensity)$_{f,t}$	**0.01495**	**0.01765** *	**0.01679** *	**0.05457** ***	**0.07130** ***	**0.07916** ***
	(1.56)	(1.80)	(1.72)	(3.29)	(4.49)	(4.95)
(Export-sales ratio)$_{f,t}$	**-0.00187**	**-0.00202**	**-0.00249**	**0.00894** ***	**0.00753** ***	**0.00705** ***
	(-0.90)	(-0.97)	(-1.19)	(3.92)	(3.44)	(3.21)
ln(Number of workers)$_{f,t}$	**0.00786** ***	**0.00786** ***	**0.00786** ***	**0.00470** ***	**0.00404** ***	**0.00408** ***
	(12.92)	(12.84)	(12.84)	(16.01)	(16.13)	(16.14)
(Non-production worker share)$_{j,t}$	**0.01740** ***	**0.03771** ***	**0.03219** ***	**0.12853** ***	**0.05089** ***	**0.04780** ***
	(4.36)	(9.79)	(8.58)	(17.58)	(13.11)	(11.78)
(Capital-labor ratio)$_{j,t}$	**0.00002** **	**0.00004** ***	**0.00002** *	**-0.00004** ***	**0.00001**	**0.00002** ***
	(1.99)	(3.64)	(1.95)	-6.57	(1.46)	(3.08)
(R&D intensity)$_{j,t}$	**0.26853** ***	**0.34714** ***	**0.32071** ***	**-0.58364** ***	**-0.33512** ***	**-0.18455** ***
	(4.15)	(5.29)	(4.92)	-17.04	-15.11	-9.04
(Export intensity)$_{j,t}$	**-0.03174** ***	**-0.02985** ***	**-0.03523** ***	**-0.00016** ***	**-0.00012** ***	**-0.00013** ***
	(-5.83)	(-5.06)	(-6.36)	(-24.95)	(-18.47)	(-22.26)
(Import intensity)$_{j,t}$	**0.00705** **	**0.01037** ***	**0.00695** **	**0.00003** ***	**0.00001** ***	**0.00001** ***
	(2.27)	(3.11)	(2.29)	(10.33)	(5.78)	(4.31)
$\ln(FDI)_{j,t}-\ln(FDI)_{j,t-1}$	**0.00505** ***			**-0.00826** ***		
	(3.27)			(-3.83)		
(FDI to China share)$_{j,t}$	**-0.05020** ***			**-0.16281** ***		
	(-11.57)			(-11.12)		
(FDI to USA share)$_{j,t}$	**-0.00096**			**0.02202** ***		
	(-0.22)			(7.68)		

Table 6. (Continued)

	Korea			Japan		
	(1)	(2)	(3)	(4)	(5)	(6)
$(\text{FDI to Japan share})_{j,t}$	**-0.07005** ***					
	(-4.74)					
$(\text{FDI to Korea share})_{j,t}$				**0.04393** ***		
				(5.42)		
$\ln(\text{Exports})_{j,t}-\ln(\text{Exports})_{j,t-1}$		**-0.02749** ***			**-0.00032**	
		(-7.13)			(-0.19)	
$(\text{Exp to China share})_{j,t}$		**-0.00942**			**0.02221** ***	
		(-1.15)			(3.23)	
$(\text{Exp to USA share})_{j,t}$		**-0.02249** ***			**0.06286** ***	
		(-2.70)			(15.05)	
$(\text{Exp to Japan share})_{j,t}$		**0.01064**				
		(1.31)				
$(\text{Exp to Korea share})_{j,t}$					**0.02929** ***	
					(5.34)	
$\ln(\text{Imports})_{j,t}-\ln(\text{Imports})_{j,t-1}$		**-0.01117** *			**-0.01674** ***	
		(-1.95)			(-12.6)	
$(\text{Imp from China share})_{j,t}$		**-0.01428** *			**-0.04698** ***	
		(-1.71)			(-14.85)	
$(\text{Imp from USA share})_{j,t}$		**-0.08476** ***			**-0.01110** ***	
		(-6.70)			(-3.31)	
$(\text{Imp from Japan share})_{j,t}$		**-0.02696** ***				
		(-4.41)				
$(\text{Imp from Korea share})_{j,t}$					**-0.01085** ***	
					(-3.22)	
$\ln(\text{Trade})_{j,t}-\ln(\text{Trade})_{j,t-1}$			**-0.04212** ***			**0.00916** ***
			(-6.42)			(4.85)
$(\text{Trade with China share})_{j,t}$			**-0.01677** **			**-0.01428** ***
			(-2.02)			(-2.59)

Table 6. (Continued)

	Korea						Japan					
	(1)		(2)		(3)		(4)		(5)		(6)	
(Trade with USA share)$_{j,t}$					**-0.05837**	***					**0.01045**	***
					(-5.99)						(2.63)	
(Trade with Japan share)$_{j,t}$					**-0.04081**	***						
					(-5.46)							
(Trade with Korea share)$_{j,t}$											**-0.01472**	*
											(-1.67)	
Intercept	**0.09893**	***	**0.10265**	***	**0.10006**	***	**-0.08903**	***	**-0.06419**	***	**-0.06079**	***
	(7.03)		(7.24)		(7.08)		(-28.11)		(-23.15)		(-25.47)	
Number of observations	30,076		29,475		29,475		29,077		37,354		37,354	
F value		***		***		***	193.7	***	201.6	***	210.2	***
R-squared	0.3974		0.3971		0.3954		0.3253		0.3068		0.2953	

Note: Heteroskedasticity-robust t-statistics in parentheses.

In previous studies, Ito (2005) found a negative correlation between exposure to imports from low-wage countries and employment growth at manufacturing firms in Japan, while Ahn *et al.* (2005) suggested that outbound FDI to China has a negative impact on employment growth and productivity growth in Korean manufacturing establishments. This paper, applying fixed-effect panel regressions to Korean and Japanese micro-data, provides a more nuanced perspective on the earlier findings. According to the results of Ahn *et al.* (2005), which used pooled ordinary least square (OLS) regression with industry and year dummies, plant-level average employment growth and TFP growth in Korea tend to be lower in industries with a higher share of outward FDI to China. However, these negative effects of transactions with China on employment and productivity growth in Korea's manufacturing sector disappear once plant-specific fixed effects are controlled for. However, in Japan, the shares of FDI to and trade with China still have a significant negative impact, particularly on the firm-level TFP growth rate, even after controlling for firm-specific fixed effects. The negative impact of outsourcing to China thus is more pronounced in Japan than in Korea, which may reflect the fact that Japanese firms are more advanced in outsourcing to China than their Korean counterparts. Further research paying closer attention to the link between the evolutions of global production networks in East Asia and their impact on regional economies would be necessary for deeper understanding of the different results for Korea and Japan.

Another intriguing finding is that the industry-level R&D intensity has a positive effect on TFP growth in Korea but a negative one in Japan. A possible explanation of this difference in the contribution of industry R&D to the productivity growth of individual plants/firms is the distance to the technological frontier. As most Korean manufacturing sectors still have not reached the technological frontier, it is relatively easy for them to imitate and absorb more advanced technologies. Such spillover effects would explain the positive contribution of industry-level R&D intensity. On the other hand, most Japanese manufacturing sectors are very close to the technological frontier. Consequently, technology is more sophisticated and cannot be imitated quickly. Near the technological frontier, technological competition among firms is fierce and the competition effect is likely to dominate the spillover effect. This is an issue that deserves further attention. More rigorous empirical studies based on micro-data for East Asian economies could help to shed more light on the development process of firms' capabilities and firms' research and development strategies.

APPENDIX: OECD CLASSIFICATION OF MANUFACTURING BY TECHNOLOGY INTENSITY (1997)

High-technology Industries

- Aircraft and spacecraft
- Pharmaceuticals
- Office, accounting and computing machinery
- Radio, TV and communications equipment

- Medical, precision and optical instruments

Medium-high-technology Industries

- Electrical machinery and apparatus, n.e.c.
- Motor vehicles, trailers and semi-trailers
- Chemicals excluding pharmaceuticals
- Railroad equipment and transport equipment, n.e.c.
- Machinery and equipment, n.e.c.

Medium-low-technology Industries

- Coke, refined petroleum products and nuclear fuel
- Rubber and plastic products
- Other non-metallic mineral products
- Building and repairing of ships and boats
- Basic metals
- Fabricated metal products, except machinery and equipment

Low-technology Industries

- Manufacturing, n.e.c., and recycling
- Wood, pulp, paper, paper products, printing and publishing
- Food products, beverages and tobacco
- Textiles, textile products, leather and footwear

Appendix Table 1. Description of Variables

Explanatory Variables	Definition
Plant(firm)-Level Variables	
(Sales growth, 3-year)$_{f,t}$	Growth rate of shipment (sales); 3-year average from year *t* to year *t+3*
ln(TFP)$_{f,t}$	Logarithm of total factor productivity level in year *t*
(Non-production worker share)$_{f,t}$	Share of non-production worker in total number of workers in year *t*
(Capital-labor ratio)$_{f,t}$	Ratio of capital stock to number of workers in year *t*
(R&D intensity)$_{f,t}$	Ratio of R&D expenditure to shipment (sales) in year *t*
(Export-sales ratio)$_{f,t}$	Ratio of exports to shipment (sales) in year *t*
(Import-sales ratio)$_{f,t}$	Ratio of imports to shipment (sales) in year *t*
ln(Number of workers)$_{f,t}$	Logarithm of number of workers in year *t*

Appendix Table 1. Description of Variables

Explanatory Variables	Definition
Industry-Level Variables	
(Non-production worker share)$_{j,t}$	Share of non-production worker in total number of workers in year *t*
(Capital-labor ratio)$_{j,t}$	Ratio of capital stock to number of workers in year *t*
(R&D intensity)$_{j,t}$	Ratio of R&D expenditure to shipment (sales) in year *t*
(Export intensity)$_{j,t}$	Ratio of exports to shipment (sales) in year *t*
(Import intensity)$_{j,t}$	Ratio of imports to shipment (sales) in year *t*
ln(FDI)$_{j,t}$-ln(FDI)$_{j,t-1}$	Growth rate of outbound FDI stock from year *t-1* to *t*
(FDI to China share)$_{j,t}$	Share of FDI stock to China in total outbound FDI stock in year *t*
(FDI to USA share)$_{j,t}$	Share of FDI stock to USA in total outbound FDI stock in year *t*
(FDI to Japan share)$_{j,t}$	Share of FDI stock to Japan in total outbound FDI stock in year *t*
(FDI to Korea share)$_{j,t}$	Share of FDI stock to Korea in total outbound FDI stock in year *t*
ln(Exports)$_{j,t}$-ln(Exports)$_{j,t-1}$	Growth rate of exports from year *t-1* to *t*
(Exp to China share)$_{j,t}$	Share of exports to China in total exports in year *t*
(Exp to USA share)$_{j,t}$	Share of exports to USA in total exports in year *t*
(Exp to Japan share)$_{j,t}$	Share of exports to Japan in total exports in year *t*
(Exp to Korea share)$_{j,t}$	Share of exports to Korea in total exports in year *t*
ln(Imports)$_{j,t}$-ln(Imports)$_{j,t-1}$	Growth rate of imports from year *t-1* to *t*
(Imp from China share)$_{j,t}$	Share of imports to China in total exports in year *t*
(Imp from USA share)$_{j,t}$	Share of imports to USA in total exports in year *t*
(Imp from Japan share)$_{j,t}$	Share of imports to Japan in total exports in year *t*
(Imp from Korea share)$_{j,t}$	Share of imports to Korea in total exports in year *t*
ln(Trade)$_{j,t}$-ln(Trade)$_{j,t-1}$	Growth rate of trade (exports plus imports) from year *t-1* to *t*
(Trade with China share)$_{j,t}$	Share of trade (exports plus imports) to China in total exports in year *t*
(Trade with USA share)$_{j,t}$	Share of trade (exports plus imports) to USA in total exports in year *t*
(Trade with Japan share)$_{j,t}$	Share of trade (exports plus imports) to Japan in total exports in year *t*
(Trade with Korea share)$_{j,t}$	Share of trade (exports plus imports) to Korea in total exports in year *t*

REFERENCES

Ahn, S. H., K. Fukao, and H. U. Kwon (2004), "The Internationalization and Performance of Korean and Japanese Firms: An Empirical Analysis Based on Micro-Data," *Seoul Journal of Economics*, vol. 17, no. 4, pp. 439-82.

Ahn, S. H., D. H. Lee, S. W. Lee, and C. S. Woo (2005), "The Economic Impacts of Outbound FDI and Trade: The Case of Korea," paper presented at the *OECD Workshop on the Globalisation of Production*, 15-16 November, 2005, Paris.

Amiti M. and S. J. Wei (2006), "Service Offshoring and Productivity: Evidence from the United States," *NBER Working Paper* 11926, National Bureau of Economic Research.

Aw, B. Y., X. Chen, and M. J. Roberts (2001), "Firm-Level Evidence on Productivity Differentials, Turnover, and Exports in Taiwanese Manufacturing," *Journal of Development Economics*, vol. 66, pp. 51-86.

Aw, B. Y., S. Chung, and M. J. Roberts (2000), "Productivity and Turnover in the Export Market: Micro-Level Evidence from the Republic of Korea and Taiwan (China)," *The World Bank Economic Review*, vol. 14, no. 1, pp. 65-90.

Barba Navaretti, G. and A. J. Venables (2004), *Multinational Firms in the World Economy*, Princeton: Princeton University Press.

Barba Navaretti, G. and D. Castellani (2004), "Investments Abroad and Performance at Home: Evidence from Italian Multinationals," *CEPR Discussion Paper* No. 4284, March, Centre for Economic Policy Research.

Barba Navaretti, G., D. Castellani, and A. C. Disdier (2006), "How Does Investing in Cheap Labour Countries Affect Performance at Home? France and Italy," mimeo.

Bernard, A. B., J. B. Jensen, and P. Schott (2006), "Survival of the Best Fit: Exposure to Low-Wage Countries and the (Uneven) Growth of U.S. Manufacturing Plants," *Journal of International Economics*, vol. 68, pp. 219-237.

Blomström, M., Fors, G. and R. E. Lipsey (1997), "Foreign Direct Investment and Employment: Home Country Experience in the United States and Sweden," *Economic Journal*, vol. 107, pp. 1787-97.

Brainard, L. and D. Riker (1997), "Are U.S. Multinationals Exporting U.S. Jobs?," *NBER Working Paper* 5958, National Bureau of Economic Research.

Brainard, L. and D. Riker (2001), "Are U.S. Multinationals Exporting U.S. Jobs?," in D. Greenaway and D. R. Nelson, eds., *Globalization and Labour Markets*, Vol. 2, pp. 410-426, Elgar, Cheltenham, U.K. and Northampton, M.A.

Desai, A. M., Foley, C. F., and J. R. Hines (2005a), "Foreign Direct Investment and the Domestic Capital Stock," *NBER Working Paper* 11075, National Bureau of Economic Research.

Desai, A. M., Foley, C. F., and J. R. Hines (2005b), "Foreign Direct Investment and Domestic Economic Activity," *Harvard Business School Working Paper*, November 2005.

Feldstein, M. (1995), "The Effects of Outbound Foreign Direct Investment on the Domestic Capital Stock," in M. Feldstein, J. R. Hines, and R. G. Hubbard, eds., *The Effects of Taxation on Multinational Corporations*, University of Chicago Press, Chicago, pp. 43-63.

Freeman, R. B. and L. F. Katz (1991) "Industrial Wage and Employment Determination in an Open Economy," in J. M. Abowd and R. B. Freeman, eds., *Immigration, Trade and the Labor Market*, Chapter 8, University of Chicago Press, Chicago.

Good, H. D., Nadiri, M. I., and R. Sickles (1999), "Index Number and Factor Demand Approaches to the Estimation of Productivity," *Handbook of Applied Econometrics: Microeconometrics*, pp. 14-80.

Görg, H., A. Hanley, and E. Strobl (2005), "Productivity Effects of International Outsourcing: Evidence from Plant Level Data," mimeo, University of Nottingham.

Hahn, C. H. (2005), "Exporting and Performance of Plants: Evidence from Korean Manufacturing," in T. Ito and A. K. Rose, eds., *International Trade in East Asia*, NBER-East Asia Seminar on Economics, Volume 14, Chapter 2, pp. 53-78, University of Chicago Press, Chicago and London.

Harrison, A. E. and M. S. McMillan (2006), "Outsourcing Jobs? Multinationals and US Employment," *NBER Working Paper* 12372, National Bureau of Economic Research.

Head, K. and J. Ries (2002), "Offshore Production and Skill Upgrading by Japanese Manufacturing Firms," *Journal of International Economics*, vol. 58. no. 1, pp. 81-105.

Hijzen, A., T. Inui, and Y. Todo (2006), "The Effects of Multinational Production on Domestic Performance: Evidence from Japanese Firms," paper presented at the RIETI-21st Century COE Hi-Stat Program Workshop *Industry- and Firm-Level Total Factor Productivity and Economic Growth in Japan*, July 24, Research Institute of Economy, Trade and Industry, Tokyo.

Ito, K. (2005), "Chu-Tei-shotokukoku kara no Yunyu Kyogo-do to Kigyo Seicho: Kigyo Katsudo Kihon Chosa Kohyo Deta ni yoru Jissho Bunseki (Competition with Imports from Low- and Lower-Middle-Income Countries and Its Impact on the Growth of Japanese Companies: Empirical Analysis Based on Micro-Data of the "Basic Survey of Japanese Business Structure and Activities)," *RIETI Discussion Paper Series* 05-J-028, September, Research Institute of Economy, Trade and Industry.

Ito, K. and K. Fukao (2005a), "Physical and Human Capital Deepening and New Trade Patterns in Japan," in T. Ito and A. K. Rose, eds., *International Trade in East Asia*, NBER-East Asia Seminar on Economics, Volume 14, Chapter 1, pp. 7-49, University of Chicago Press, Chicago and London.

Ito, K. and K. Fukao (2005b), "Nihon no Sangyo-kan Sangyo-nai Kokusai Bungyo to Taigai Chokusetsu Toshi: Kokunai no Butteki Jinteki Shihon Shinka he no Eikyo (The Vertical Division of Labor and Japanese Outward FDI: The Impact on Human Capital Deepening in Japan)," *Keizai Bunseki* (The Economic Review), vol. 56, no. 4, pp. 331-47 (in Japanese).

Ito, K. and K. Fukao (2007), "Determinants of the Profitability of Japanese Manufacturing Affiliates in China and Other Regions: Does Localization of Procurement, Sales, and Management Matter?" *RIETI Discussion Paper Series* 06-E-037, January, Tokyo: Research Institute of Economy, Trade and Industry.

Kimura, F. and K. Kiyota (2006), "Exports, FDI, and Productivity: Dynamic Evidence from Japanese Firms," *MMRC Discussion Paper*, no. 69, University of Tokyo.

Ministry of Commerce, People's Republic of China (2004) *China Commerce Yearbook 2004*, Ministry of Commerce, People's Republic of China, Beijing.

Ministry of Commerce, People's Republic of China (2005) *China Commerce Yearbook 2005*, Ministry of Commerce, People's Republic of China, Beijing.

Ministry of Finance, Japan (2006), *Outward Direct Investment (Industry & Country)*, retrieved from: http://www.mof.go.jp/english/e1c008.htm. 29 March 2006.

Muendler, M. A. and S. O. Becker (2006), "Margins of Multinational Labor Substitution," *IZA Discussion Paper* 2131, May, Institute for the Study of Labor.

OECD (1997), *Science, Technology and Industry: Scoreboard of Indicators*, Organisation for Economic Co-Operation and Development, Paris.

Paprzycki, R. and K. Fukao (2005), "The Extent and History of Foreign Direct Investment in Japan," *Hi-Stat Discussion Paper Series*, No. 84, April, Institute of Economic Research, Hitotsubashi University.

Revenga, A. L. (1992), "Exporting Jobs? The Impact of Import Competition on Employment and Wages in U.S. Manufacturing," *Quarterly Journal of Economics*, vol. 107, no. 1, pp. 255-284.

Riker, D. A. and S. L. Brainard (1997), "U.S. Multinationals and Competition from Low Wage Countries," *NBER Working Paper* 5959, National Bureau of Economic Analysis.

RIETI (2006), *JIP Database 2006 Provisional Version*, Research Institute of Economy, Trade and Industry, retrieved from http://www.rieti.go.jp/jp/database/d04.html. 22 May 2006.

Sachs, J. D. and H. J. Shatz (1994) "Trade and Jobs in U.S. Manufacturing," *Brookings Papers on Economic Activity*, vol. 19, pp. 1-84.

Statistics Bureau, Japan (various years) *Establishment and Enterprise Census*, Ministry of Internal Affairs and Communications, Japan.

In: Micro-Evidence for the Dynamics of Industrial Evolution ISBN: 978-1-60456-388-7
Editors: A. Heshmati, J.-D. Lee

Chapter 5

THE GLOBALIZATION OF JAPANESE FIRMS AND ITS IMPACT ON THE DOMESTIC ECONOMY

Kazuyuki Motohashi*
Department of Technology Management for Innovation
University of Tokyo and Research Institute of Economy, Trade and Industry,
Tokyo, Japan

ABSTRACT

In an era of globalization, it is natural for large firms to seek attractive markets and cost advantage by producing in low labor cost countries. However, large firms shifting of their production sites overseas may have a negative impact on SMEs, which do not have enough management resources for overseas production. This is particularly the case for SMEs whose major customer has shifted its production sites.

In this paper, the impact of the globalization of large firms on the domestic economy, particularly on the business activities of SMEs, is analyzed by using firm level data of METI's BSBSA (Basic Survey of Business Structure and Activity) and SOBA (Survey of Overseas business Activities). It is found that there is no direct evidence of negative impact of globalization on SMEs in this study. More and more SMEs are becomng involved in global production activities, and a complementary relationship between domestic and overseas sales is observed.

---

* Professor, Department of Technology Management for Innovation, School of Engineering, University of Tokyo, e-mail: motohashi@tmi.t.u-tokyo.ac.jp. Views expressed in this paper are those of author's, and not those of his organization. This paper is based on preliminary result of RIETI's research project on competitiveness of Japanese industry. Author would like to thank for Toshiyuki Matsuura (RIETI) for his inputs to constructing panel datasets of BSBSA and SOBA.

Introduction

In the postwar years, the Japanese economy caught up with the United States and Europe at an astonishing speed. Japanese automobiles and electronic products spread worldwide, and in the 1980s, even Made in America, a Massachusetts Institute of Technology (MIT) report on U.S. industrial competitiveness, rated the practices of Japan's auto and semiconductor industries as the best in the world (Dertouzos, M.L. et al (1989)). In the late 1980s, the Japanese economy entered its longest postwar expansionary phase. This economic boom, later to be described as the "bubble" economy, ended abruptly in the early 1990s following the 1990 stock market crash and the subsequent tumble in land prices. In the wake of the collapse of the so-called bubble, the 1980s approbation of Japan's industrial competitiveness became the economic gloom of the 1990s.

Facing a sluggish domestic market, Japanese firms started looking at international markets more seriously. In addition, Japanese firms have to rely on low cost labor in developed countries, in order to meet intense international market competition. All of these factors contribute to the growing trend of globalization activities in Japan. For large multinational companies, globalization means the reallocation of their production facilities across borders. However, structural changes caused by increasing overseas production may have non negligible impacts on domestic production and employment.

In this paper, the impact of the globalization of large firms on the domestic economy, particularly on the business activities of SMEs, is analyzed, by using firm level data of METI's BSBSA (Basic Survey of Business Structure and Activity) and SOBA (Survey of Overseas business Activities). The relationship between globalization and domestic production depends on the type of overseas production, and the objectives of globalization have been changing over time and regions. Such microstructure of overseas production activities, as well as their impact on SME's business activities is analyzed.

The structure of paper is as follows. The next section provides am overview of globalization activities of Japanese firms. SOBA's statistics of overseas production is compared with trade statistics to understand the size of Japanese multinationals' activities, focusing on electronics industry. In section 3, possible impacts of globalization on domestic economy are discussed. In addition, descriptive statistics comparing global and non-global firms are presented. Then, a section for economic analysis follows. In this section, the relationship between domestic and overseas production is analyzed by econometric models. Finally, this paper concludes with a summary and the next step of this study.

1. Overview of the Global Activities of Japanese Firms

In an era of globalization, firm's activities across national borders such as international trade and investment become more and more important. In the 1980s, Japan's manufacturing firms, particularly in electronics and the automotive industry, enjoyed a strong presence in the world market. Growing exporting shares in world trade triggered political concern in the US and European countries, and a series of trade negotiations were initiated. After the economic bubble burst in the early 1990s, the Japanese economy entered into a period of slow economic growth, and the importance of international markets has never been as small since.

Reaching an international market can be done not only by export, but also overseas production activities. Japanese firms have increased their overseas activities by foreign direct investments since the mid 1980s, when the Japanese yen appreciated substantially due to the 1985 Plaza Accord. Firstly, Japanese manufacturers expanded their production facilities in South East Asian countries such as Thailand and Malaysia. After the 1990s, many firms set up their production sites in China, backed by China's open door policies.

In this section, global activities by Japanese firms are reviewed by input-output tables, trade data and overseas activity statistics. First, Figure 1 shows the impact of international trade on the domestic economy by comparing the export shares (the share of exports in total production) and the import penetration ratios (the share of imports in total domestic demand) between 1990 and 2004, based on the data from input-output tables. It is interesting to see that both export share and import penetration ratio have increased in all industries. In some industries, such as textiles, apparel and non-ferocious metals, import penetration ratio is relatively greater. In these industries, Japanese manufacturers are not so competitive in the global market. On the other hand, in general machinery and transportation machinery, the export share is relatively larger, which implies that Japanese firms have a stronger position in the world.

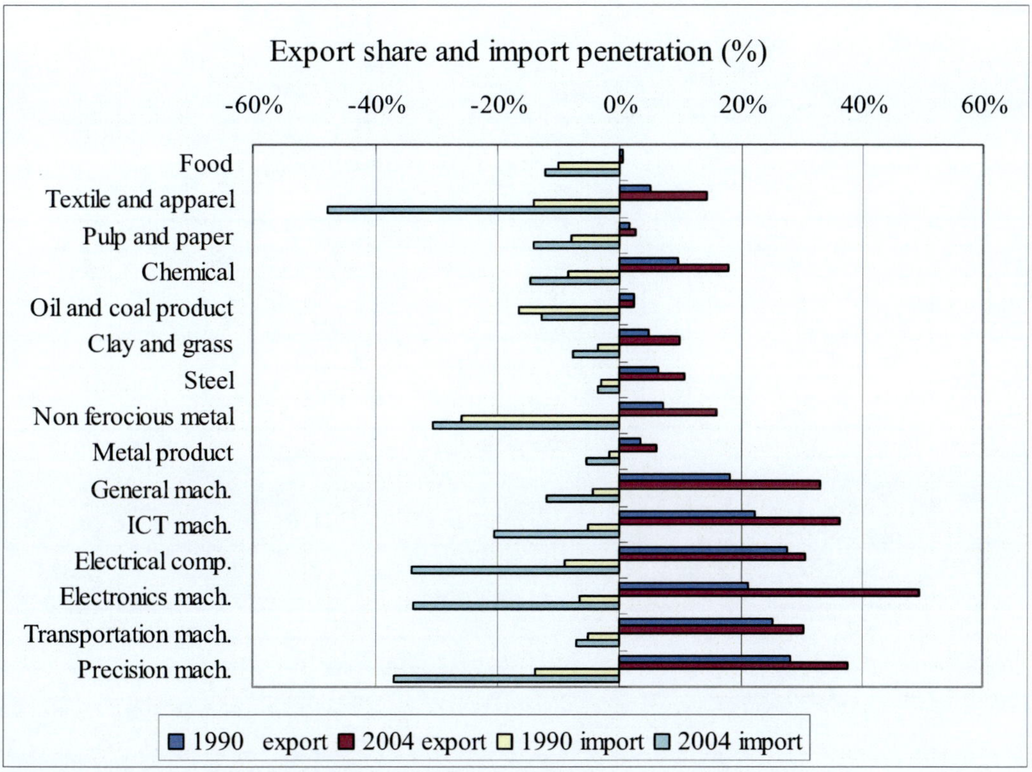

Source: Input Output Tables (Statistical Bureau and METI).

Figure 1. Export share and import penetration ratio by industry.

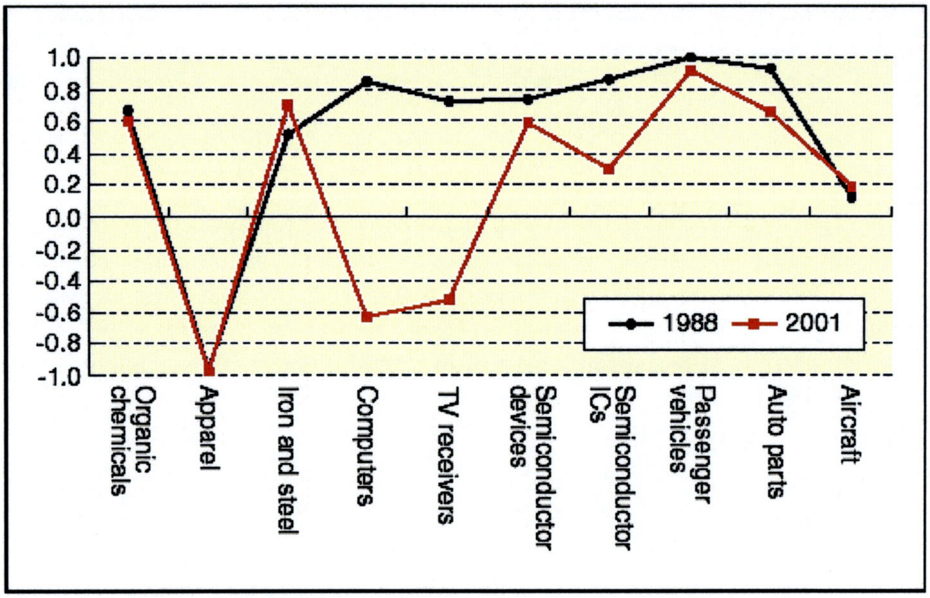

Source: **Author's calculation using "Trade Statistics," Ministry of Finance**

Source: Author's calculation using "Trade Statistics." Ministsry of Finance.

Figure 2. Trade specialization index for Japan-East Asia trade.

Table 1. Trade statistics by regions

	ICT Machinery							
	Import				Export			
	Total (billion yen)	NIES3	China	ASEAN4	Total (billion yen)	NIES3	China	ASEAN4
1990	704	21.9%	3.8%	4.5%	5470	8.4%	6.9%	1.9%
2005	3370	12.9%	52.8%	18.1%	3250	9.0%	10.2%	2.1%

	Electronics Machinery							
	Import				Export			
	Total (billion yen)	NIES3	China	ASEAN4	Total (billion yen)	NIES3	China	ASEAN4
1990	345	27.3%	9.6%	6.6%	1580	17.4%	6.7%	8.5%
2005	1130	15.7%	45.7%	14.9%	2230	13.0%	22.0%	11.6%

	Electronics Components							
	Import				Export			
	Total (billion yen)	NIES3	China	ASEAN4	Total (billion yen)	NIES3	China	ASEAN4
1990	764	25.9%	1.7%	7.2%	3430	30.6%	8.2%	9.2%
2005	3700	35.4%	22.2%	19.9%	7980	25.7%	33.8%	14.2%

Source: Trade Statistics (MOF).

It should be noted that both export share and import penetration ratio are relatively high in the electronics industries.[1] In addition, the import penetration ratio increased faster during the period from 1990 to 2004 than the export share in these industries, which may be due to a reduction of the international competitiveness of the Japanese electronics industries. In order to further investigate this point, Figure 2 shows trade specialization indexes (TSIs)[2] by commodity level trade statistics. TSI is a typical indicator to assess the international competitiveness of an industry. Figure 2 compares TSIs in 1988 and 2001 for trade with East Asia. Notable changes of TSIs in this period can be found in electronics industries. For example, in final demand products such as computers and TV receivers, export surpluses have turned into import surpluses. Although changes are much smaller for electronic components such as semiconductor devices and ICs, downward trends of TSI can be found. These findings suggest that the relative competitive position of Japanese firms went down. Alternatively, the impact of globalization on the domestic economy may be greater for the electronics industry than for other industries, because a substantial portion of international trade in these sectors can be explained by the overseas activities of Japanese firms.

Table 1 shows the trade statistics of electronic industries by trading partners. Here, we focus on Asian countries, because of the increasing importance of these countries as a trading counterpart. First, it is found that imports grow faster than exports, and more than half the imports of electronics industries come from Asian countries. Second, within Asian countries, the share of China and ASEAN increases, while that of NIES3 decreases. Particularly, it is noticeable that the share of China is increasing very fast as an import counterpart. Finally, the exports of components industries expand faster than those of machinery industries (ICT and electronics machinery), and the share of Asian countries as exporting countries is relatively smaller than that of importing countries.

A growing international linkage with Asian countries can also be found in the overseas activities of Japanese firms. In Figure 3, the trends of overseas production of three electronics sectors is presented, using the data from Survey of Overseas Business Activities (SOBA). In electronics machinery and components sectors, more than half the overseas production is located in Asian countries. It is also found that the share of production in China is increasing. In contrast, the share of Asian countries in ICT machinery sector is relatively small, while it is gradually increasing.

In Table 2, the amount of overseas production is compared with that of domestic production for three electronics sectors. From 1990 to 2004, the domestic production of ICT and electronics machinery sectors decreased, while that of the electrical components sector increased a little. A dramatic shift from domestic to overseas production is found in the ICT machinery sector. In 1990, the amount of overseas production used to be less than half the domestic production, while the size of overseas production is now surpassing that of the domestic production in recent years. In general, a machinery producer tends to relocate its production activities to overseas sites as compared with a component producer. In a machinery producer, there are a lot of assembling activities, which are typically labor intensive jobs. Therefore, relocating such activities to Asian countries with relatively cheap

---

[1] Here, electronics industries are broken up into three sub-categories, ICT machinery (including computers and communications equipments), Electronics components and electronics machinery (including consumer electronics and industrial electronics machinery).

[2] TSI is defined as the ratio of the value of trade surplus (export – import) to the value of total trade (export + import).

labor is economically reasonable. On the other hand, producing electrical components involves a lot of capital intensive work, which can be found in semiconductor factories as a typical example.

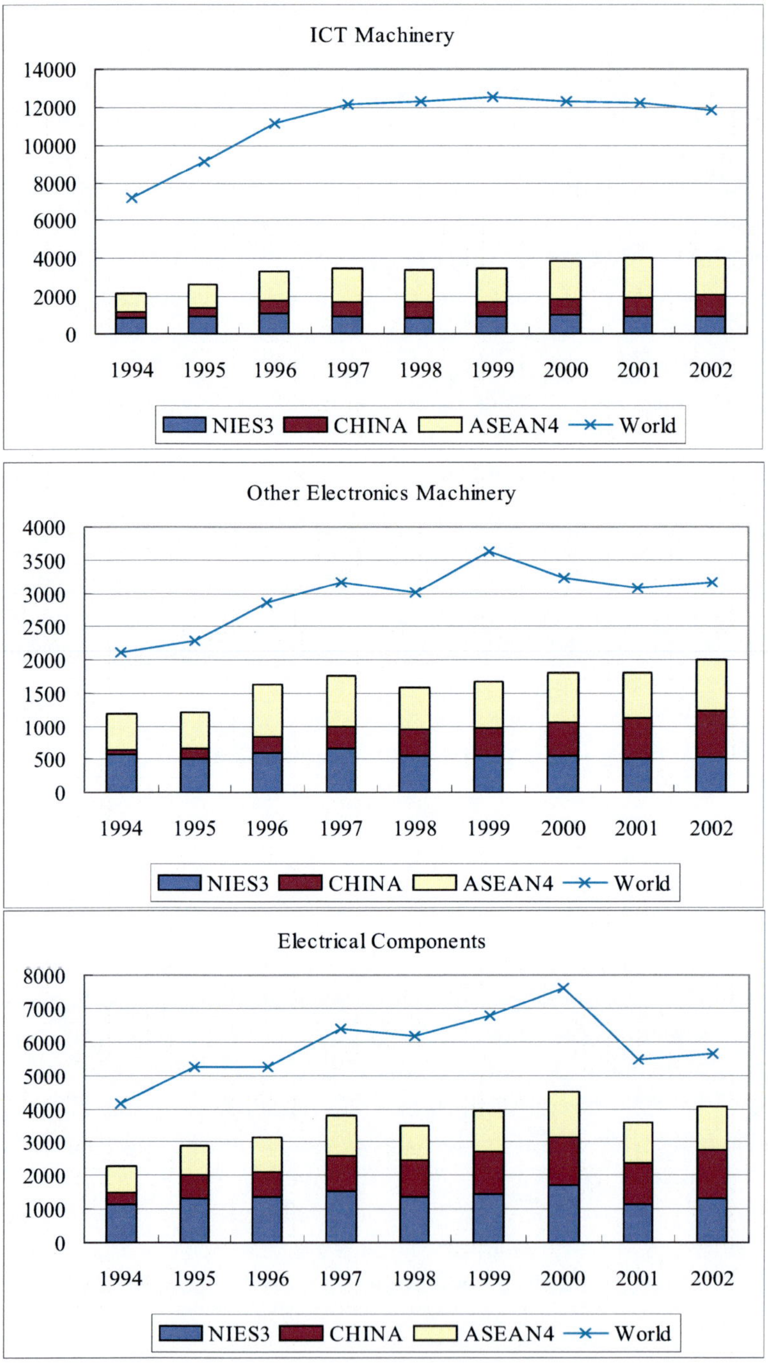

Figure 3. Overseas Production of Electronics Industries (Unit: billion yen).

Table 2. Comparing Domestic and Overseas Production (Unit: billion yen)

	1990				2004			
	Prod.	Export	Import	Overseas Prod.	Prod.	Export	Import	Overseas Prod.(*)
ICT machinery	12,082	3,342	848	5,551	8,904	2,710	3,214	11,833
Electronics machinery	26,524	5,857	1,126	1,992	20,201	7,313	3,319	3,160
Electrical components	12,169	2,564	661	2,455	13,329	6,575	3,428	5,632

Note: Figures in 2002.
Source: Survey of Overseas Business Activities (METI).

This difference in production location patterns between machinery and component producers affects the trade pattern of these products. In Figure 3, a simple diagram is presented for understanding the impact of overseas production on trade statistics. In this framework, there are two sectors: assembly and components. These two sectors used to be located inside one country, but the assembly sector has relocated its production to an overseas site. We assume that both domestic demand (Da) and overseas demand (Oa) do not change. Now the products made outside the country serve not only overseas demand (Oa), but also part of the domestic demand (Da'), taking advantage of low cost overseas production. At the same time, the amount of exports, which used to be domestic products serving overseas demand (Oa), disappears because overseas demand can now be met by the overseas production site. Therefore, the exports of the assembly sector decrease (Oa->0), and its imports increase (0->Da'). In addition, the amount of domestic production decreases by the amount of overseas production (Da'+Oa).

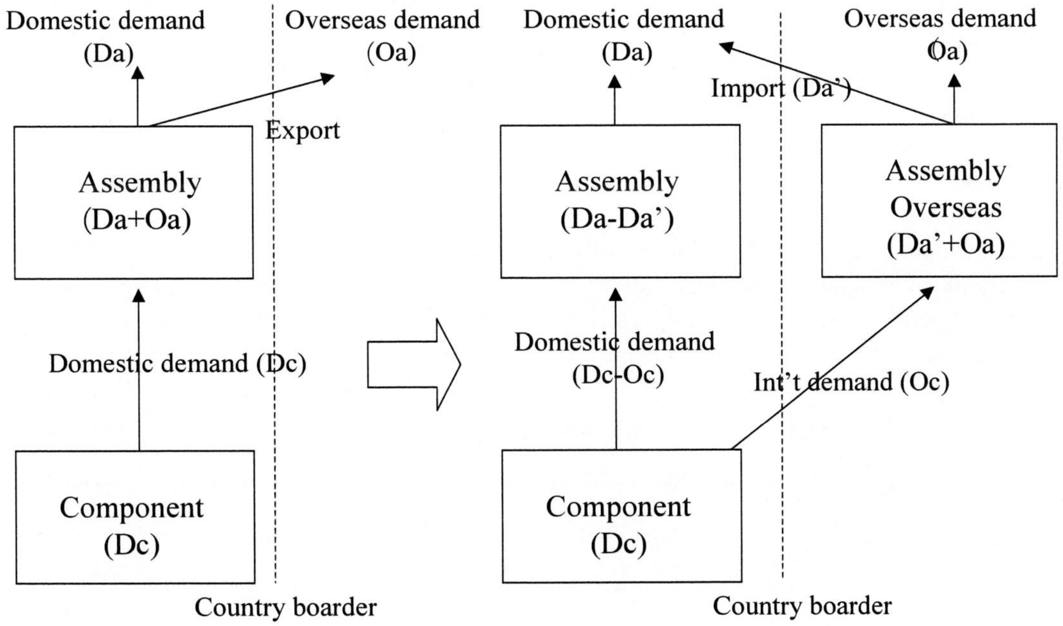

Figure 4. Overseas production by assembly firms.

The impact of the components sector is completely different from that of the assembly sector. We assume that this sector does not move its production site overseas, and supplies its products only to this assembly sector. By overseas relocation of the assembly sector, some of components have to be exported to service the demand from the overseas production site by the assembly sector. Therefore, the exports of the component sector increase (0->Oc'). However, the import and the production levels do not change under this framework. These patterns of the assembly and component sectors are basically consistent with those which are presented in Table 2.

In Table 3, the impact of overseas production on trade is presented for three electronics sectors. First, a large proportion of imports occurring in the assembly sector (ICT and electronics machinery here) can be explained by the importing back of overseas production to Japan. The figure of the component sector (electrical component here) is a little smaller than that of the assembly sector. There are no comparable statistics for export, because the data of overseas purchases from Japan (i.e., exports derived by overseas production) is aggregated by the sector of output, not by the type of products used in that production. For example, the overseas ICT machinery sector purchased 2,714 billion yen from Japan in 2002. A substantial proportion of this amount is electronics components used in overseas production, but there may also be others. However, when we compare the size of purchases in all electronics sectors (2,714+1,221+679=4,614) to the amount of electronics component exports (6,280), the impact of overseas production on the amount of exports of this sector is supposed to be non-negligible.

Japanese companies in the electronics industry have been transferring their production operations, particularly assembly processes, abroad to East Asia. This globalization of business activities by Japanese companies is basically in line with the Heckscher-Ohlin model of international trade theory, i.e., global location of production sites depending on factor endowments. By relocating labor intensive work to China and ASEAN countries, with abundant and cheap labor, Japanese manufacturers can achieve a higher level of efficiency in their production process.

However, what is happening in reality is more complicated than what is predicted by classical trade theory. Figure 5 shows the export share and import penetration ratio for electronics goods, by the most detailed level of commodity classification of 2000 input and output table. It is found that these two indices had a positive relationship in 2000, which contradicts classical trade theory's comparative advantage story. There is a some literature to show a growing share of intra-industry trade (Fukao et, al, 2003; Miyagawa et. al, 2003). Figure 5 shows this trend can be observed even in the detailed level of industry classification.

Table 3. Impact of Overseas Production on Trade Statsitics (Unit: billion yen)

	Import	Import by overseas prod.	(Share)	Export	Export by overseas puchase
ICT machinery	2,910	1,386	(47.6%)	3,840	2,714
Electronics machinery	925	343	(37.1%)	1,840	679
Electrical components	2,760	906	(32.8%)	6,280	1,221

Source: Survey of Overseas Business Activities (METI) and Trade Statistics (MOF).

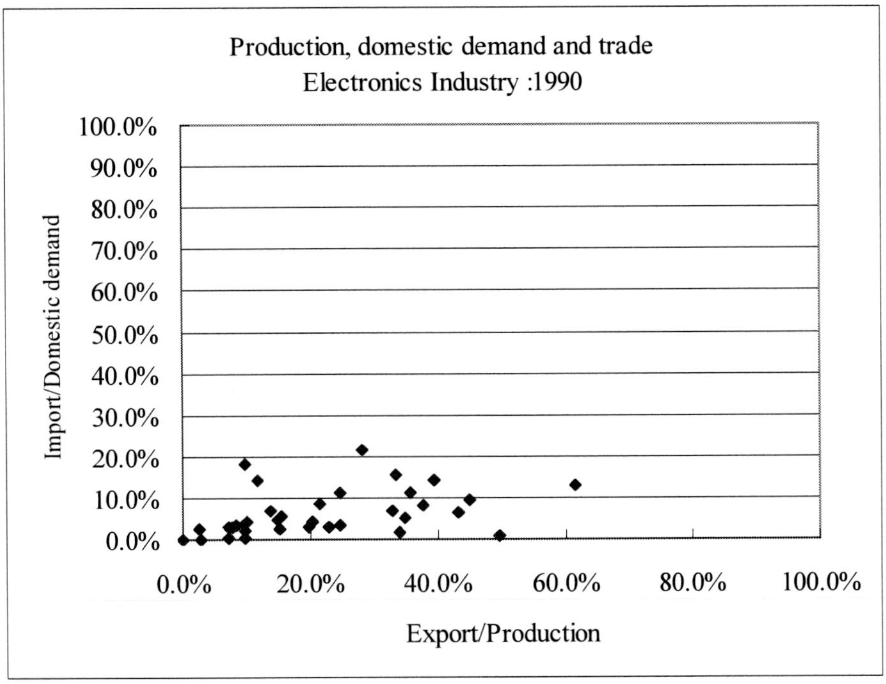

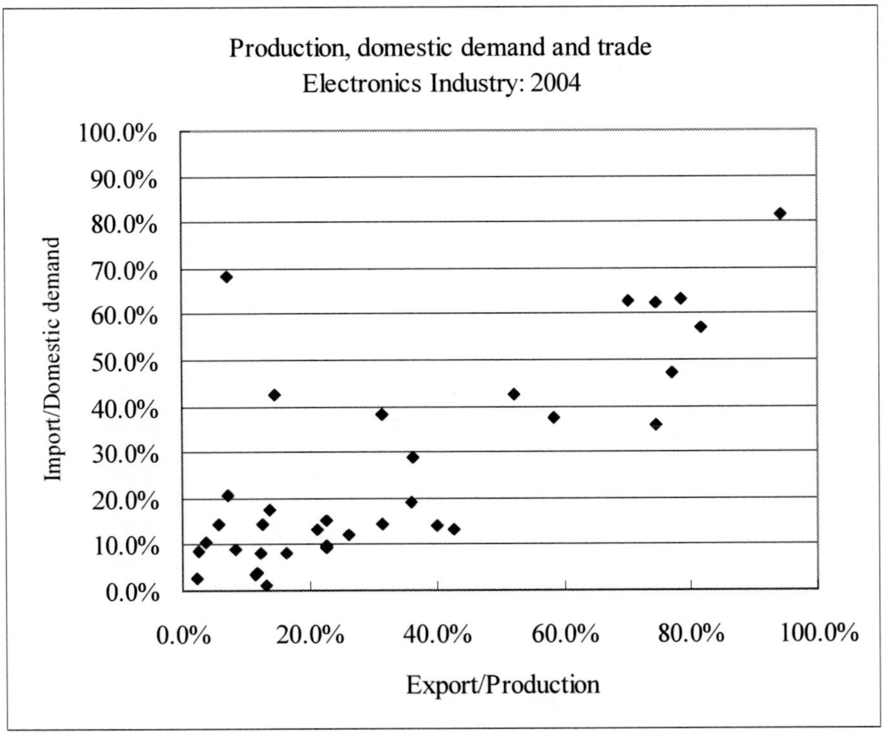

Figure 5. Export Share and Import Penetration Ratio by Detail Commodity Classification.

Growing intra industry trade suggests an increasing degree of differentiation in each commodity, reflecting increasing complexity of product architecture of electronics goods.[3] For example, the semiconductor chip is now used in almost all products. The semiconductor chip used by mobile phones is totally different from that used by digital TVs, and optimal allocation of different types of semiconductor induces trade within this very narrowly defined industry. In addition, modulalization of product architecture, particularly the case for electronics goods, leads to fragmentation of production activities across borders.[4] The PC is the most obvious illustration: many vendors supply many varieties of, for example, hard-disk drives, but any can be plugged into the computer. Modularization of product architecture leads to unbundling the supply chain, and dividing it among various kinds of players. Advancement of information technology reduces communication costs, which further pushes the geographical dispersion of production activities of one product (Motohashi, 2006).

2. Globalization's Impact on the Domestic Economy

There are numerous studies on the economic impact of globalization. In terms of structural changes associated with globalization, many studies focus on the impact of the labor market. As is found in the previous section, one of the major motivations for the globalization of Japanese firms is to seek low labor costst. In this case, presumably, they relocate unskilled workers' jobs overseas, and focus on skilled workers' jobs in the home country. This leads to decreasing demand for unskilled labor in the home countries, and the difference in wages between skilled and unskilled workers becomes wider (Feenstra, 1996: Slaughter 1995). In Japan, Sakurai (2000) addresses this issue by using industry level data. Higuchi and Genda (1999) also show the positive relationship between employment in subsidiaries in developed countries and the share of white collar workers in parent companies in Japan.

However, it is not always the case that increasing globalization activities have a negative impact on domestic production and employment. For example, Bloomstrom et. al (1997) show that the FDIs of US firms in developed countries have positive effects on domestic employment, while those in developing countries have negative ones. Chen and Ku (2003) also find a complementary relationship between domestic production and overseas production in Taiwanese multinational companies. Multinationals may be able to gain managerial expertise by accessing the international market, and this will particularly be the case for FDI in developed countries.

These studies suggest that a firm's globalization activity takes various forms, and it is important to take into account its heterogeneity in analyzing economic impacts. Economic implications may be different, not only for the host country, but also for various other factors such as the organizational form of foreign subsidiaries, the degree of intra firm transactions and backward linkage with local suppliers. Kimura and Ando (2004) show detailed descriptive statistics for Japanese multinationals in order to understand the state of

[3] New trade theory, taking into account product differentiation and increasing returns of scale, supports the existence of intra-industry trade by using a theoretical model (Helpman and Krugman, 1985)

[4] Modular architecture refers to the design of a product (mostly physical, but also some software) using standardized interfaces to allow the interchangeability of various specific varieties of a component, and the combining of components into what can be complex systems. (Baldwin and Clark, 2002)

'international fragmentation of production sites'. For example, it is found that the share of intra-firm transactions is decreasing for both sales and procurement of foreign subsidiaries in East Asia. This point can be confirmed by Kiyota et. al. (2005) showing that experience in the local market is one of the important determinant factors of backward linkage with local suppliers.

In terms of globalization's impact on the domestic economy, there is political concern about "hollowing out". A large multinational company can relocate its production site overseas, by taking into account factor costs and the market size of the host country. However, it may be difficult for small companies to globalize their production facilities due to the lack of managerial resources. In this case, moving out of production activities from Japan induces substantial structural adjustments in particular sectors and regions. In this section, possible impacts of globalization on the domestic market are reviewed by focusing on Japanese SMEs in manufacturing sector.

In order to look at the impact of globalization, the data from METI's BSBSA (Basic Survey of Business Structure and Activities) are used in the following part. BSBSA is a firm level census for manufacturing and wholesale/retail sector with no less than 50 employees or 30 million yen capital amount. This survey started in 1991, and an annual survey has been conducted since 1994. Survey items cover a wide range of a firm's structures and activities, such as employment by function, sales by industry, innovation activities, globalization and the ownership structure.

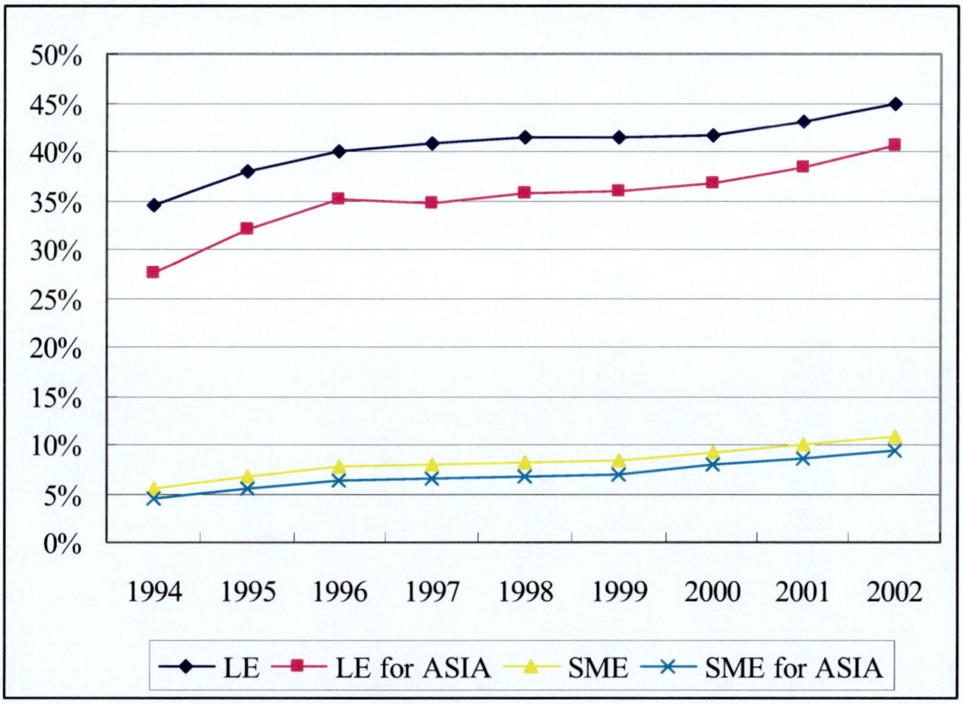

Source: Author's calculation based on BSBSA.

Figure 6. Share of firms with overseas production.

In Figure 6, the globalization activities of SMEs and large firms (LEs) are compared. For both types of firms, the shares of firms with overseas production in any region and those in Asian countries are presented. Globalization activities are increasing for both LEs and SMEs, although the share of SMEs with overseas production is still small. In addition, it is found that most of the SMEs overseas production activities take place in Asia. Presumably, a major objective for SMEs overseas production is to cut production costs, instead of seeking new overseas markets.

This point can be confirmed by Figure 7, which is based on the results of an opinion survey for globalizing companies (METI, 2004). This graph indicates the share of firms responding 'relevant' to each type of local factors in their FDI decision. As for 'labor costs', more than half the firms indicated 'relevant' for all types of firms. In contrast, the share of 'relevant' for 'market size' decreases as the firm size gets smaller. In addition, the share of 'infrastructure' and 'legal and tax environment' is smaller for small firms as well. It may be the case that small firms are forced to set up overseas production sites for any reason such as the necessity of cost reduction and major customer's requirements.

Within the Asian region, the objectives of overseas production are different across countries. Figure 8 shows the results of an opinion survey for SME White Paper again. In China and NIES countries, the major motivation for overseas production is 'cheap product import back' as compared to 'responding to Japanese customer's needs'. In contrast, the share of 'overseas market' and 'respond to Japanese customer's needs' is greater than that of 'cheap product import back' in south east Asia. This is due to the fact that a substantial number of Japanese large firms have constructed supply chain networks in South East Asian countries, and ASEAN are promoting such investments by intra region free trade agreements (AFTA) and an investment treaty (AIA).

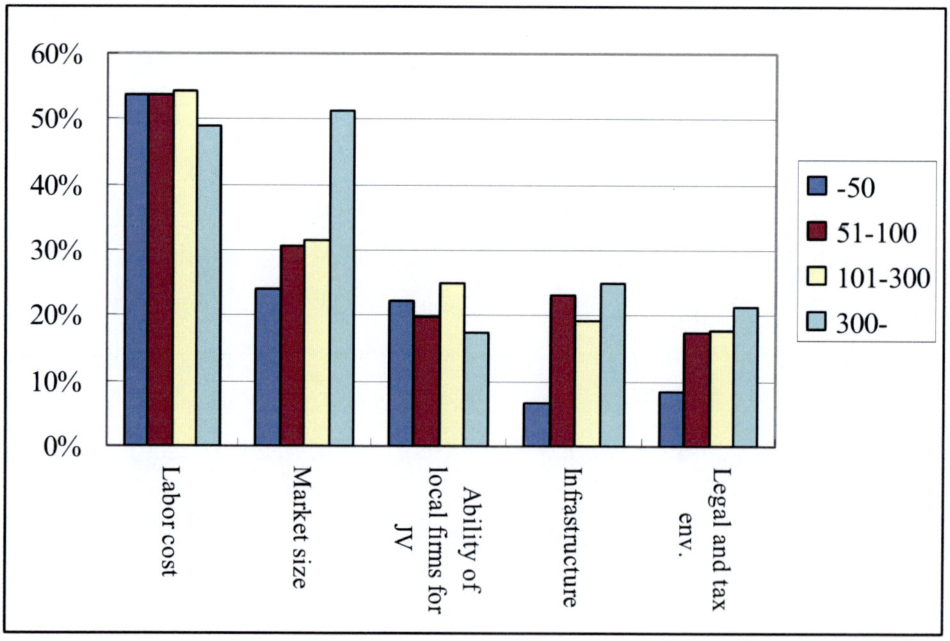

Source: 2004 SME White Paper in Japan (METI).

Figure 7. Relevant local factors for FDI.

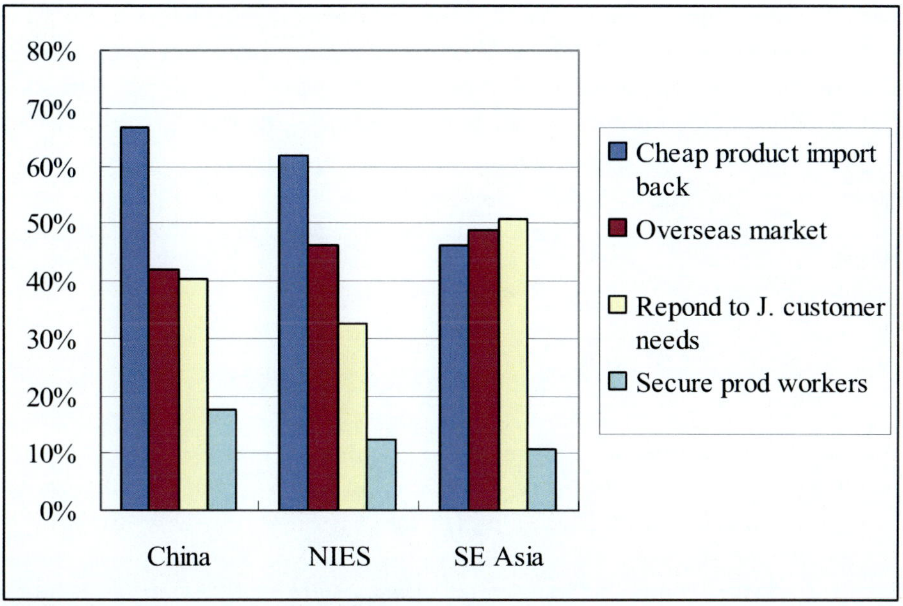

Source: 2004 SME White Paper in Japan (METI).

Figure 8. Objectives of overseas production by location.

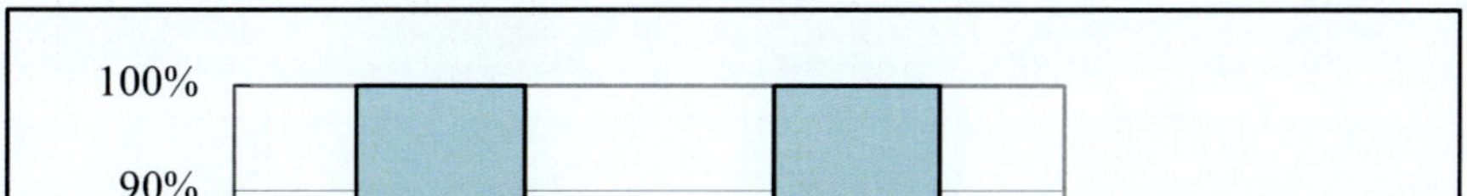

Source: Author's calculation based on BSBSA.

Figure 9. Sales share by firm size and globalization.

Figure 9 shows the share of sales cross classified by firm size and by overseas production. The share of large firms with overseas production (Global LE) increases and reached more than 60% in 2002. In contrast, the share of large firms without overseas production (Domestic LE) becomes smaller. As for SMEs, the share of firms with overseas production (Global SMEs) increases, but its sales share is still very small. Although the share of global SMEs increases, the share of domestic SMEs does not decline. Therefore, the shift of output from domestic firms to global firms can be seen in large firms, but this is not the case in SMEs.

3. ECONOMETRIC ANALYSIS

In this section, a quantitative analysis of the relationship between domestic and overseas production is presented. The dataset used in this section is a linked data of BSBSA used in the previous section with data from Survey of Overseas Business Activity (SOBA) by METI. SOBA is an annual survey conducted since 1971 for all Japanese firms with foreign subsidiaries.[5] A survey instrument is sent to parent companies located in Japan, and each parent company is supposed to answer all questions concerning its foreign subsidiaries. The sample size in the most recent survey in the 2003 fiscal year is about 4,000 firms, and MITI received the data from 2,411 firms (response rate: 64.9%) with 13,856 foreign subsidiaries. In order to investigate the relationship between domestic and overseas production, the data from SOBA for overseas production have to be linked with BSBSA with various parent company's data. In this study, a panel dataset from 1994 to 2002 for all firms with domestic and international manufacturing activities is constructed with around 1,000 samples for every year.[6]

Using this dataset, the amount of domestic sales is regressed by that of overseas subsidiaries as well as other factors at firm level. For both BSBSA and SOBA, activity base information (sales by commodity and services) is available, and the data for only manufacturing activities, instead of the total amount of figures of the firm, is generated. Therefore, the amount of manufacturing sales is a good measure of the size of production activities, and regression results will show how each firm group allocates production activities between domestic and foreign production sites. Explanatory variables for log of domestic sales are as follows,

- LOS: Log of sales in overseas subsidiaries

[5] A foreign subsidiary is defined as a company no less than 10% of whose stocks are owned by a parent company in Japan, or a so-called grandchild company of no less than 50% owned by a child company of no less than 50% owned. All parent companies in Japan, except those in financial services and real estate sector are covered.

[6] Sample size for each year is as follows,

year	# of firms
1994	871
1995	852
1996	881
1997	1,055
1998	1,029
1999	1,081
2000	1,049
2001	918
2002	842
Total	8578

- D-lf: Dummy variable for parent firm with more than 300 employees (large firm dummy)
- LOS*D-lf: Cross term of LOS and D-lf
- LRD: Log of R&D in parent company
- D_Parent: Dummy variable for whether parent company is a subsidiary of another company

Table 4 shows regression results by fixed effect model estimation. LRD is included in this regression model in order to control for time variant unobservable factors for a firm's success (in domestic sales), which cannot be controlled by a fixed effect model.

One of the major findings is that overseas production does not substitute for domestic production, but rather these two are in a complementary relationship according to positive and statistically significant coefficients to LOS. This result is consistent with those in Chen and Ku (2003) for Taiwan, it can be interpreted that overseas production can strengthen the competitive position of whole firm groups, which leads to an expansion of domestic sales as well. This is the case not only for overseas production in developed countries, but also for that in developing countries such as China.

Positive coefficients to D-lf show that domestic sales in large firms grow faster than those of SMEs. More interestingly, negative and statistical significant coefficients can be found in cross terms of LOS and D-lf for overseas production in Asia, implying that the degree of substitution between domestic and international sales is greater for large firms. In the case of China, the effect of LOS on domestic sales can be described as $\frac{\partial \ln(domsales)}{\partial LOS} = 0.061 - 0.061 * Dlf$. Therefore, a positive impact of LOS is offset by Dlf in case of large firms (Dlf=1). It should be noted that this cross term effect can be observed only in Asia, and no difference between large firms and SMEs is found in North America and Europe.

Table 4. Regression Results-1 (Dependent Variable=Log(domestic sales))

	North America	Europe	NIES	ASEAN	China	Others
LOS	0.055	0.060	0.083	0.067	0.061	-0.029
	(2.69)**	(1.91)	(3.70)**	(3.43)**	(2.83)**	(0.87)
D-lf	0.508	0.366	0.590	0.527	0.661	-0.071
	(3.31)**	(1.79)	(3.62)**	(3.64)**	(4.21)**	(0.35)
LOS*D-lf	-0.039	-0.035	-0.056	-0.045	-0.061	0.046
	(1.89)	(1.13)	(2.42)*	(2.25)*	(2.73)**	(1.35)
LRD	0.055	0.065	0.046	0.044	0.041	0.025
	(6.98)**	(6.64)**	(5.45)**	(5.68)**	(3.40)**	(2.09)*
D_Parent	-0.030	-0.067	0.007	-0.093	-0.042	-0.193
	(1.00)	(1.51)	(0.19)	(2.75)**	(0.73)	(2.44)*
Constant	10.037	10.391	9.797	9.932	10.015	11.575
	(62.77)**	(47.09)**	(58.96)**	(65.82)**	(57.16)**	(51.81)**
Observations	3768	2359	2987	3913	2413	1319
Number of kikatsu	717	457	573	759	533	256
R-squared	0.03	0.04	0.03	0.03	0.02	0.02

Absolute value of t statistics in parentheses.

* significant at 5%; ** significant at 1%.

Table 5. Regression Results-2 (Dependent Variable=Log(domestic sales))

	NIES		ASEAN		China	
	(1)	(2)	(3)	(4)	(5)	(6)
LOS	0.097	0.095	0.068	0.073	0.072	0.074
	(3.72)**	(3.65)**	(3.31)**	(3.48)**	(3.17)**	(3.26)**
D-lf	0.663	0.711	0.509	0.567	0.745	0.753
	(3.48)**	(3.70)**	(3.33)**	(3.31)**	(4.54)**	(4.48)**
LOS*D-lf	-0.066	-0.065	-0.041	-0.047	-0.072	-0.075
	(2.42)*	(2.39)*	(1.98)*	(2.17)*	(3.09)**	(3.19)**
%SAJ	0.005	-0.092	-0.008	-0.088	0.024	-0.063
	(0.13)	(0.92)	(0.26)	(0.99)	(0.64)	(0.64)
%SUJ	0.021	0.171	0.032	0.152	0.066	0.161
	(0.79)	(2.15)*	(1.43)	(1.92)	(2.47)*	(2.21)*
%SAJ*D_lf		0.117		0.087		0.098
		(1.11)		(0.96)		(0.97)
%SUJ*D_lf		-0.167		-0.127		-0.108
		(2.03)*		-1.58		-1.41
LRD	0.048	0.047	0.046	0.046	0.041	0.041
	(5.20)**	(5.12)**	(5.49)**	(5.50)**	(3.25)**	(3.21)**
D_Parent	0.007	0.006	-0.098	-0.100	-0.063	-0.062
	(0.16)	(0.16)	(2.66)**	(2.71)**	(0.99)	-0.98
Constant	9.532	9.501	9.782	9.726	9.802	9.797
	(49.62)**	(49.13)**	(61.46)**	(55.85)**	(53.66)**	(52.93)**
Observations	2561	2561	3423	3423	2057	2057
Number of kikatsu	552	552	734	734	515	515
R-squared	0.04	0.04	0.03	0.03	0.03	0.03

Absolute value of t statistics in parentheses.
* significant at 5%; ** significant at 1%.

In order to further investigate the linkage of domestic and overseas production, the following explanatory variables are added.

- %SAJ: Share of overseas sales import back to Japan to total overseas sales in overseas subsidiaries
- %SUJ: Share of supply of parts and materials to overseas production from Japan to total amount of supply
- %SAJ*D_lf: cross term of %SAJ and D_lf
- %SAJ*D_lf: cross term of %SAJ and D_lf

Fixed effect estimation results for Asian countries are presented in Table 5.

Negative coefficients are expected to %SAJ, because import back sales of overseas production will compete with domestic production. Conversely, positive coefficients to %SUJ can be expected, because this stimulates the export of domestically produced products. From regression results in Table 4, some statistically significant results can be found in %SUJ, but not in %SAJ. Although some of coefficients are not at a statistically significant level, the sign of coefficients show that there is a positive relationship between domestic sales and the export of domestic parts and materials to overseas production. But this impact is smaller for large firms, according to the coefficients to %SUJ*D_lf. In contrast, the import back of overseas production does not reduce domestic sales. This supports the hypothesis of gaining competitiveness by overseas production in the sense that globalization stimulates domestic production more than the size of imports back of overseas production.

The foregoing discussion is based on the regression results for globalization firms, and it is found that globalization does not have a negative impact on domestic activities in general, but it rather has a positive impact, particularly for SMEs. Then the next question is what is happening for non globalization firms. Figure 10 shows the changes in the globalization ratio (the share of global firms in total sales) and the changes in the SME ratio (the share of SMEs in total sales) by industry, based on BSBSA data. The industrial classification is based on the 3 digit industry classification with 58 manufacturing sectors. It can be shown that a negative relationship between these two indicators, i.e., as the globalization ratio increases, the SME ratio decreases.

This point is further investigated by using linked dataset of BSBSA and SOBA. Industry level data for domestic and overseas sales is constructed. In order to match the industrial classification of BSBSA and SOBA, the number of industries reduces to 35, but this data is cross classified by with or without overseas production, region of overseas production, employment category, and with or without parent company, as well. In total, over 400 observations are available in each year from 1994 to 2002. Table 6 shows the regression results of the log of domestic sales as dependent variables, by the random effect model. Regressions are conducted separately for 4 groups, i.e non glonalization SME, non globalization LE, globalization SME and glonalization LE.

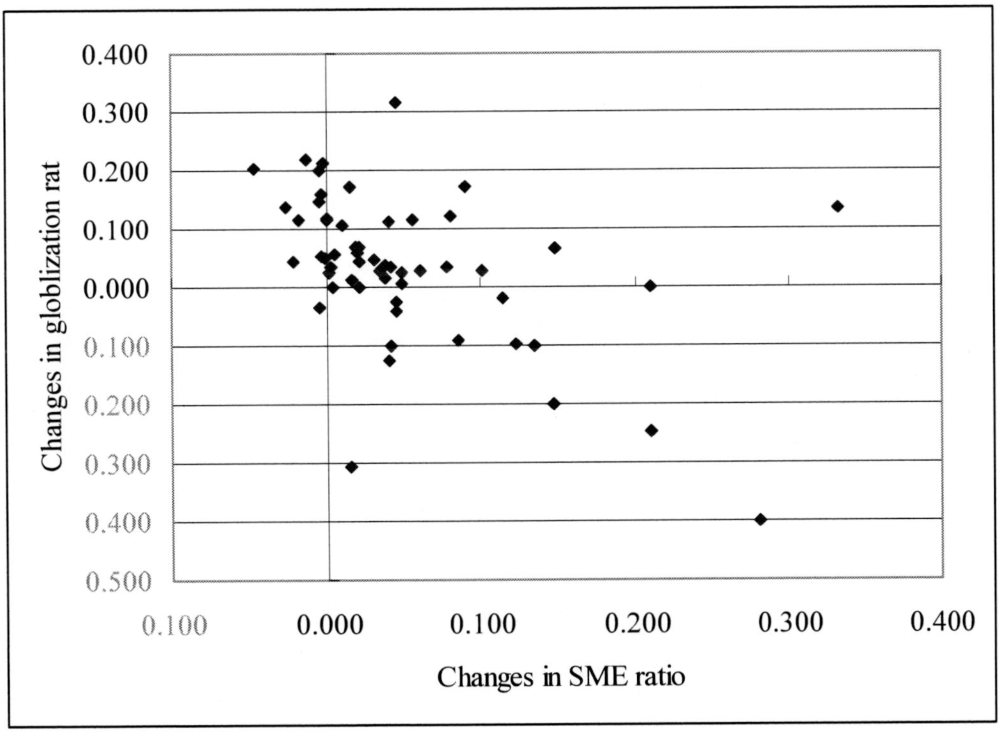

Source: Author's calculation based on BSBSA.

Figure 10. Globalization and SME by industry.

Table 6. Regression Results-3 (Dependent Variable=Log(domestic sales))

	Overseas=No		Overseas=Yes	
	SME	LE	SME	LE
LOS	-0.016	-0.035	0.043	0.026
	(1.17)	(1.86)+	(2.46)*	(1.72)+
LRD	0.461	0.524	0.339	0.546
	(32.11)**	(26.25)**	(19.94)**	(40.42)**
D_parent	-0.438	-1.307	-0.822	-0.463
	(2.48)*	(4.95)**	(2.85)**	(1.96)+
LOS*D_parent	0.022	0.117	0.011	-0.017
	(1.34)	(4.92)**	(0.42)	(0.83)
Constant	9.303	8.686	8.267	8.032
	(44.12)**	(30.27)**	(37.13)**	(35.27)**
Observations	1000	930	705	830
Number of ind	35	35	35	34
R-squared	0.47	0.53	0.48	0.83

Absolute value of t statistics in parentheses.
+ significant at 10%, * significant at 5%; ** significant at 1%.

A clear contrast between the globalization and non globalization group can be found, i.e., a negative impact of overseas production in the non globalization group, and a positive one for the globalization group. However, the negative impact of overseas production on non globalization firms is small when it is a subsidiary of another firm, according to positive coefficient to cross term of LOS and D_parent. A coefficient of LOS for non globalized SME is negative, but not statistically significant, in contrast to that for non globalized LE. Therefore, it can be said that the negative impact of globalization on the domestic economy is smaller for SMEs.

4. Conclusion

In this paper, the economic implications of globalizing industrial activities by Japanese firms are analyzed and discussed. It is observed that overseas production activities of Japanese firms have been increasing steadily since the early 1990's. Growing overseas production, particularly in the electronics industries, has a significant impact on trade statistics. In addition, it is found that the share of intra-firm trade increases even in a very detailed industry classification. Modularization of electronics product architecture and decreasing communication costs induce fragmentation of production activities across country borders.

However, major players in overseas activities are large multinationals, and the SME's position is still limited, even though its rate of growth is very high. Foreign direct investments involve significant business and political risks in the host country, and a large amount of capital investment is needed. In this sense, it is natural that such activities are concentrated in large firms. Therefore, there may be a concern over globalization's negative impacts on SMEs which do not have enough management resources for overseas production.

However, we cannot find direct evidence of a negative impact of globalization on SMEs in this study. More and more SMEs are becomng involved in global production activities, and a complementary relationship between domestic and overseas sales is observed. As compared to globalization firms, the growth rate of domestic sales is smaller for non globalization firms. Some negative impacts of globalization on these firms is found, but this is particularly the case for large firms, instead of SMEs.

Hitomi and Motohashi (2004) provide detailed information on how SMEs respond to competition from international markets, by an interview survey of 49 Japanese firms. There are some firms which shifted their production sites to China or South East Asian countries to seek lower cost production, but, there are other firms which stayed in Japan and compete against international competition by upgrading manufacturing technology. There are also other SMEs, focusing on product design, and keeping high profit level as a fab-less company. In the U.S., plant level adjustment to relative product demand changes induced by international competition has been analyzed in detail by Bernard et. al (2006). The next step of this study is to conduct the same type of systematic analysis of Japanese firms' responses to globalization, based on more detail datasets, such as manufacturing census data, to see the representative-ness of the case studies in Hitomi and Motohashi (2004).

REFERENCES

Baldwin, C. and K. Clark. (2000), *Design Rules, The Power of Modularity*, MIT Pressm Cambridge, MA.

Bernart, A., Jensen, B. and P. K. Schott (2006), Survival of the best fit: Exposure to low-wage countries and the (uneven) growth of US manufacturing plants, *Journal of International Economics*, vol. 68, pp. 219-227.

Broomstrom, T., Fors, G. and R. lipsey (1997), Foreign Direct Investment and Emoloyment: Home Country Experience in the United States and Sweden, *Economic Journal*, 107(445), pp. 1787-97.

Chen, T-J and T-H Ku (2003), The Effect of Overseas Investment on Domestic Employment, NBER Working Paper 10156.

Dertouzos, M.L. et al (1989), *Made in America*, MIT Press, Cambridge MA.IMD (2002), *World Competitiveness Yearbook 2002.*

Feenstra, R. (1996), Foreign Investment, Outsourcing and Relative Wage, in R. Feenstra and G. Douglas eds, *Political Economy of Trade Policy: Essays in Honor of Jagdish Bhagwati*, Cambridge MA: MIT Press, pp. 89-127.

Fukao, K., Ishido, M., Ito, K. and Y. Yoshiike (2003), Vertical Intra-Industry Trade and Foreign Direct Investment in East Asia, RIETI Discussion Paper 03-J-009, 2003/06, RIETI Tokyo (in Japanese).

Higuchi, Y. and Y. Genda (1999), Globalization of manufacturing SMEs and its impact on labor market, in Sekiguchi and Higuchi ed., *Industry and Employment in Global Era*, Toyo Keizai Shimpo-sha (in Japanese).

Hitomi, K and K. Motohashi (2004), Survey on Global Location Strategy of Product Base in Manufacturing Sector, Socio-Economic Research Center Rep No. Y03019, CRIEPI, Tokyo, Japan (in Japanese).

Kimura, F. and M. Ando (2004), Two-dimensional Fragmentation in East Asia: Conceptual Framework and Empirics, Forthcoming in *International Review of Economics and Finance.*

Kiyota, K. Matsuura, T., Urata. S. and Y Wei (2005) , Reconsidering the Backward Vertical Linkage of Foreign Affiliates: Evidence from Japanese Multinationals, RIETI Discussion Paper Series 05-E-019.

METI (2004), Small Medium Enterprises White Paper 2004, Ministry of Economy, Trade and Industry, Tokyo, Japan.

Miyagawa, T, Takeuchi, B. and T. Yoshikawa (2003), Are Asian Countries Threats to Japan? Changes in Trade Structure and International Competitiveness, *JCER Review,* vol. 52, October 2003 (in Japanese).

Motohashi, K. (2006), The IT Revolution's Implications for the Japanese Economy, in *Japan: Moving Toward a More Advanced Knowledge Economy*, T. Shibata ed., World Bank Institute, Washington DC, 2006/06.

Sakurai, K. (2000), Globalization and Labor Market, *Keizai Keiei kenkyu*, September 2000, Japan Development Bank (in Japanese).

Slaughter, M. (1995), Multinational Corporations, Outsourcing and American Wage Diversion, *NBER Working Paper 5253.*

In: Micro-Evidence for the Dynamics of Industrial Evolution ISBN: 978-1-60456-388-7
Editors: A. Heshmati, J.-D. Lee

Chapter 6

INTERNATIONAL INTRAFIRM TRANSFER OF MANAGEMENT TECHNOLOGY BY JAPANESE MULTINATIONAL CORPORATIONS*

Shujiro Urata, Toshiyuki Matsuura*† *and Yuhong Wei
Waseda University; Research Institute of Economy, Trade and Industry; and Hitotsubashi University, Tokyo, Japan

ABSTRACT

The authors analyze the extent of international intrafirm transfer of management technology achieved by Japanese manufacturing MNCs and identify its determinants by using firm-level data. Results show that top management has not been transferred to MNCs' affiliates, but labor management has. Management technology has been transferred at European affiliates but not much at ASEAN affiliates. The length of operation of the affiliates and the quality of labor are found to have positive impacts on technology transfer for the Asian affiliates. These findings indicate the importance of providing FDI friendly environment and improving labor quality to promote intrafirm transfer of management technology.

Keywords: international intrafirm transfer; management technology; Japanese multinational corporations

---

* [Reprinted from: Ureta S., T. Matsuura and Y. Wei (2007), "International Intrafirm Transfer of Management Technology by Japanese Multinaional Corporations", in Heshmati A., Y-B. Sohn and Y-R. Kim, Eds. "Commercialization and Transfer of Technology: Major Country Case Studies", Chapter 8, Nova Science Publishers.

† Corresponding author: 1-3-1 Kasumigaseki, Chiyoda-ku, Tokyo 100-8901 Japan, Tel: +81-3-3501-8412 E-mail: matsuura-toshiyuki@rieti.go.jp

1. INTRODUCTION

Technological progress plays an important role in promoting economic growth, as it enables an economy to expand even if the factors of production such as labor and capital remain at the same level. Indeed, technological progress is considered necessary to maintain and accelerate economic growth, as the increase in the factors of production is likely to terminate sooner or later.

Technology is a rather broad concept, ranging from production techniques to management know-how. As will be shown below in the literature review, many previous studies on international technology transfer focused on the technology related to production. Despite its importance on improvement of productivity of a firm, management know-how has not been studied much in the context of achieving economic growth and technology transfer in the past, partly because of the difficulty in quantifying management know-how.

International technology transfer takes various modes including purchasing technology from foreign countries in various forms including licensing, importing published materials, inviting foreign experts, foreign direct investment (FDI), and others. Among these modes of international technology transfer, FDI has become a major mode of international technology transfer in recent decades. Several reasons may be found for such developments. One is rapid expansion of FDI in the world, which in turn has been due to liberalization in FDI policies and the substantial reduction in the costs of international communication. These two factors contributed to rapid expansion of FDI as they reduced the costs of undertaking FDI. Another reason that FDI has become a major mode of international technology transfer has to do with the strategy of multinational corporations (MNCs). Recognizing the importance of keeping technology inside MNCs to maintain their competitiveness, MNCs have been rather reluctant in selling their technologies to other firms in the form of licensing.

Two types of technology transfer involving MNCs can be identified. One is technology transfer from parent firms of MNCs to their overseas affiliates, and the other is technology transfer from overseas affiliates of MNCs to local firms. The former type of technology transfer is characterized as intrafirm technology transfer, the latter as technology spillover.[1] Intrafirm technology transfer has to take place before technology spillover is realized.

Intrafirm technology transfer is carried out by various means, including work experience (on the job training), and training programs to local employees. Technology spillover may be realized in different forms. Technology may be transferred from foreign firms to local firms, when local workers who have acquired knowledge from working at foreign firms move to local firms or start new business. Local firms may acquire technology from foreign firms by learning production and management technology or know-how from their business engagement with foreign firms through procurement of parts and components or sales of products, and by imitating production methods and management know-how practiced by foreign firms.

The objective of this paper is to analyze the extent of international intrafirm transfer of management technology achieved by Japanese manufacturing MNCs and to identify the determining factors that could explain the extent of technology transfer. Our choice of

---

[1] More precisely, technology spillover is defined as technology transfer, which is undertaken without market transactions. See Saggi (1999) for a survey of the literature on technology spillover and Javorcik and Spatareanu (2005) for a difficulty investigating the presence or absence of technology spillover.

Japanese MNCs was dictated by the availability of the data. We examine the extent of technology transfer for eight different job classifications, chief executive officer (CEO), deputy chief executive officer, chief of the sections including labor management, accounting, procurement, sales, research and development (R&D), and corporate planning.

An analysis of international intrafirm transfer of management technology would prove important not only for the host countries of FDI but also for MNCs. Host countries are concerned about the outcome of intrafirm technology transfer because successful technology transfer improves the technological capability of local workers, thereby contributing to economic growth. Indeed, host governments as well as employees working at the affiliates of foreign firms have often expressed dissatisfaction with the slow pace of technology transfer by MNCs. MNCs pay a lot of attention to intrafirm technology transfer because the performance of overseas affiliates of MNCs depends crucially on the success of intrafirm technology transfer, which would lead to efficient production and management.

The structure of the remainder of the paper is as follows. Section 1 presents a brief review of previous studies on international intrafirm technology transfer, in order to set the stage for our analysis. Although most past studies examined international intrafirm transfer of technologies related to production or R&D, and not management know-how, a review of past studies should prove useful for deepening our understanding of the issue. Section 2 examines the extent of international intrafrim transfer of management technology achieved at overseas affiliates of Japanese MNCs. Section 3 has two subsections; one presents the hypotheses to explain the extent of technology transfer achieved and the other analyzes the results of the regression analysis of the determinants of the extent of technology transfer. Section 4 concludes the paper.

2. A Brief Review of the Studies on Intrafirm Technology Transfer

Several studies have examined the patterns of intrafirm technology transfer from parent firms to their overseas affiliates. Most of these studies examined the resources or the costs expended for intrafirm technology transfer by utilizing information obtained from case studies. Davies (1977) studied 119 cases of technology transfer by British companies in India. He found that British companies expend more resources for technology transfer, in the form of providing such tangibles as designs and components, as well as sending personnel, to their joint ventures with Indian firms than to local Indian firms.

Based on the information about the resource costs associated with twenty-six technology transfer projects undertaken by U.S. firms in chemicals and petroleum refining and machinery, Teece (1977) found that the costs of technology transfer were higher when technology recipients were joint ventures than when they were wholly owned foreign subsidiaries. He also found that the costs were higher when technology suppliers were less experienced in technology transfer and when recipients were less experienced in manufacturing.

Ramachandran (1993) found a similar relation between equity ownership and the resources used for technology transfer in his study of the characteristics of technology transfer agreements signed by Indian firms and MNCs from the United States, United

Kingdom, and Western Europe. Analyzing the data aggregated into fourteen industries, he found that MNCs spent more resources, in the form of sending engineers and training local employees in the MNCs' home countries, for technology transfer involving wholly owned subsidiaries than in the case of joint ventures, while they spent the least resources in the case of technology transfer to independent firms. In addition, R&D by licensees was found to reduce the amount of resources spent for technology transfer, indicating that high technological capability of the technology recipient facilitates technology transfer.

Wakasugi (1996) adopted a similar approach to study the costs of technology transfer by Japanese firms. Using information on resources expended for intrafirm technology transfer for 104 firms, Wasasugi performed statistical analyses to discern the determinants of the costs and lengths of time required for transferring technology. Similar to the findings of other studies, he found that the greater the equity participation by the parent firm, the more resources spent for technology transfer. Past experience in technology transfer was found to lower the costs of technology transfer. The level of technology to be transferred was found to affect the costs of technology transfer, in that transferring high technology tends to cost more.

Although a very important issue regarding intrafirm technology transfer is to identify the circumstances and environments in which technology can effectively be transferred, the previous studies discussed above did not address this issue directly. They instead examined the costs or resources involved in technology transfer. However, costs or resources spent for technology transfer do not indicate the extent of technology transfer achieved. An increase in resources expended for technology transfer does not realize technology transfer if the resources are spent wastefully. To deal with this problem, several researchers adopted different approaches to measure the extent of technology transfer achieved.

Yamashita (1991) measured the extent of technology transfer achieved by obtaining the information on the evaluation by personnel involved in technology transfer. He asked the personnel about his or her view on the level of technology transfer achieved. Specifically, he asked to what extent, such as 50 percent or 8o percent, the expectation of technology transfer has been achieved. He found that for many affiliates simple technology such as maintenance or repair of production line has been transferred but sophisticated technologies such as developing new products have not been transferred. One problem of this approach is the subjectivity of the evaluation. A manager in charge of technology transfer is likely to give high evaluation, while a personnel actually engaged in production may give a low evaluation.

Urata (1999) adopted a different approach. He evaluated the extent of technology transfer achieved by assessing who, either staff from the parent firm or local staff, has responsibility for managing technologies or assignments. Technology transfer is deemed to have been achieved if local staff is in charge of managing technologies. Using the sample of 133 cases of intrafirm technology transfer by Japanese MNCs to their Asian affiliates, he found a positive correlation between the extent of technology transfer and the degree of equity holding by the parent company only in the case where the technologies involved are simple, such as those related to the maintenance of machines. The opposite relation was found when the technologies involved were sophisticated, such as design technologies. His interpretation was that Japanese MNCs are reluctant to transfer technologies to their foreign affiliates, and they transfer these technologies under pressure from local joint venture partners. Urata also found that technology transfer is successfully carried out when Japanese MNCs adopt measures specifically intended to promote technology transfer, such as providing manuals in the local language and holding seminars in local areas.

Several studies have examine the issue of staffing, either expatriate (from the parent company) or local staffing, at overseas affiliates of Japanese firms, an issue taken up by Urata, from business perspectives. Basically two lines of thoughts have been proposed to explain the expatriate staffing of the CEO of overseas affiliates. One emphasizes the importance of control of an affiliate by a parent company and coordination between a parent company and its affiliate, and the other places an importance on learning local business conditions. Based on these observations, Belderbos and Heijltjes (2005) investigated the determinants of expatriate staffing by Japanese electronics affiliates in Asia and found that both perspectives, that is, control-coordination and learning, were important determinants of expatriate staffing. Specifically, being consistent with the control-coordination line of thought, the probability of expatriate staffing is found high for the affiliates with low local sales orientation, high equity holdings by parent company, and large size. Supporting the learning hypothesis, experiences in local market is found to reduce the probability of expatriate staffing. In addition, they found that inter-firm linkages, or keiretsu network, also influenced expatriate staffing in that keiretsu affiliates can share information and experiences so that they do not have to rely much on expatriates.

3. Intrafirm Transfer of Management Technology by Japanese Firms

We measure the extent of technology transfer achieved by identifying the nationality of personnel in charge of given tasks or positions. If local staff, rather than Japanese staff, is in charge, it is assumed that the technology has been transferred. This methodology is admittedly very simplistic and possibly inaccurate, but in the absence of workable methodology largely because of a lack of necessary information, our methodology may be justified.

The METI survey asked Japanese MNCs whether local staff or Japanese staff were responsible for various job classifications at their overseas affiliates. We computed the proportion of the number of affiliates indicating that local staffs are responsible in total number of firms for eight job classifications and the results are shown in Tables 1 and 2.[2]

Wide variations can be found for the pattern of technology transfer achieved, or the localization of staff for different job classifications as well as for the affiliates in different regions. Technology transfer has been limited for the senior positions such as CEOs and deputy CEOs. In 2001 approximately 10-30 percent of the overseas affiliates had local personnel as CEO, while the proportions are somewhat higher around 30-50 percent for deputy CEOs. Among different job classifications, labor management registered high proportion of local staff at around 60-75 percent. This finding reflects the difficulty in managing local workers by Japanese staff because of the differences in social, cultural, and historic backgrounds and views between local and Japanese staff. The extent of technology

[2] The MITI survey collects information from overseas affiliates of Japanese firms, which are set up both in the forms of greenfield investment and mergers and acquisitions (M&As). We limited our sample to those set up through greenfield investments, because technology transfer patterns or staffing practices at the affiliates set up through M&As appears to differ substantially from those set up through greeenfield investment.

transfer for other job categories is similar as approximately 40-60 percent of the affiliates had local staff responsible for the job.

Table 1. Intrafirm Technology Transfer for Japanese Firms

	Year of Survey	ASEAN	NIES	China	Europe	North America
CEO	1995	0.259	0.280	0.351	0.194	0.199
	1998	0.196	0.213	0.265	0.242	0.167
	2001	0.143	0.186	0.193	0.289	0.190
Deputy CEO	1995	0.231	0.345	0.514	0.412	0.378
	1998	0.283	0.411	0.501	0.464	0.374
	2001	0.308	0.412	0.512	0.493	0.432
Labor Management	1995	0.724	0.567	0.813	0.634	0.603
	1998	0.695	0.606	0.734	0.700	0.691
	2001	0.696	0.579	0.716	0.730	0.648
Accounting	1995	0.627	0.582	0.645	0.527	0.508
	1998	0.558	0.593	0.589	0.580	0.568
	2001	0.548	0.558	0.610	0.610	0.551
Sales	1995	0.440	0.476	0.691	0.559	0.573
	1998	0.460	0.491	0.596	0.612	0.558
	2001	0.449	0.475	0.584	0.658	0.610
Purchase	1995	0.480	0.514	0.729	0.610	0.628
	1998	0.554	0.553	0.666	0.661	0.676
	2001	0.559	0.558	0.642	0.712	0.678
R&D	1995	0.486	0.660	0.643	0.683	0.541
	1998	0.459	0.565	0.531	0.631	0.553
	2001	0.395	0.673	0.436	0.663	0.520
Planning	1995	0.500	0.608	0.658	0.694	0.460
	1998	0.506	0.541	0.593	0.595	0.577
	2001	0.480	0.534	0.579	0.614	0.599

Note: The figures indicate the proportion of overseas affiliates of Japanese firms with local staff taking responsibility in total number of affiliates.

Source: METI, Kaigai Jigyokatsudo Kihon Chosa.

[Comprehensive Survey of Overseas Activities of Japanese Firms], various issues.

Some regional differences can be identified for the pattern of technology transfer. Technology transfer has been limited for the affiliates in ASEAN, while it has been most advanced for those in Europe. One may conjecture that the affiliates in ASEAN face difficulty in recruiting highly capable workers for the job.

Various factors are responsible for the determination of the extent of technology transfer achieved, as we will examine later. However, at this point, we would like to see the relationship between the length of operation and technology transfer, because Japanese firms tend to develop the capabilities of local workers through on-the-job training, which is given more or less continuously and constantly for the extended time period. Table 2 shows the extent of technology transfer by the length of operation. The figures in the table indicate that the extent of technology transfer increases with the length of operation for the affiliates in

Asia, especially for those in ASEAN and China, while such pattern cannot be found for the affiliates in Europe or in North America. We will examine if such pattern can still be observed by controlling the impacts of other factory by conducting statistical analysis later.

Table 2. Intrafirm Technology Transfer by the Length of Operation: 2001

		All industry				
	Length of Operation	ASEAN	NIES	China	Europe	North America
CEO	0-5	0.081	0.163	0.179	0.257	0.186
	6-10	0.143	0.139	0.191	0.243	0.224
	11-15	0.113	0.246	0.286	0.279	0.188
	16-	0.263	0.184	0.167	0.326	0.181
Deputy CEO	0-5	0.256	0.448	0.460	0.367	0.378
	6-10	0.382	0.376	0.521	0.517	0.405
	11-15	0.254	0.430	0.600	0.505	0.407
	16-	0.342	0.403	0.667	0.528	0.481
Human Resource Management	0-5	0.606	0.475	0.610	0.746	0.628
	6-10	0.719	0.506	0.734	0.683	0.548
	11-15	0.681	0.657	0.963	0.758	0.711
	16-	0.783	0.616	0.833	0.725	0.646
Accounting	0-5	0.517	0.453	0.583	0.556	0.528
	6-10	0.584	0.531	0.605	0.667	0.587
	11-15	0.531	0.651	0.786	0.642	0.548
	16-	0.554	0.552	0.667	0.589	0.551
Sales	0-5	0.403	0.543	0.522	0.684	0.557
	6-10	0.459	0.379	0.583	0.677	0.581
	11-15	0.421	0.490	0.870	0.740	0.580
	16-	0.525	0.496	0.750	0.593	0.660
Procurement	0-5	0.489	0.527	0.605	0.733	0.684
	6-10	0.572	0.455	0.632	0.741	0.634
	11-15	0.564	0.594	0.889	0.755	0.725
	16-	0.608	0.612	0.750	0.669	0.656
R&D	0-5	0.341	0.667	0.486	0.929	0.636
	6-10	0.273	0.500	0.402	0.765	0.441
	11-15	0.395	0.656	0.556	0.542	0.470
	16-	0.583	0.771	0.667	0.581	0.552
Planning	0-5	0.436	0.577	0.489	0.594	0.488
	6-10	0.387	0.444	0.582	0.767	0.548
	11-15	0.527	0.465	0.857	0.509	0.627
	16-	0.561	0.625	0.500	0.635	0.637

Note: The figures indicate the proportion of overseas affiliates of Japanese firms with local staff taking responsibility in total number of affiliates.

Source: METI, Kaigai Jigyokatsudo Kihon Chosa.

[Comprehensive Survey of Overseas Activities of Japanese Firms], No. 32, 2004.

3.1. The Determinants of Intrafirm Transfer of Management Technology: The Hypotheses

We saw the extent of intrafirm international transfer of management technology (TT) achieved by Japanese firms to their overseas affiliates in the previous section. In this section we attempt to discern the determining factors for intrafirm transfer of management technology. In the analysis we divide the possible explanatory factors into three groups. One group of factors concerns the characteristics and strategies of the Japanese parent firms, and another concerns those of their overseas affiliates. The third group of factors is related to the characteristics of the host countries of overseas affiliates of Japanese MNCs. We discuss these factors in turn below.

To begin with the characteristics of the parent firms, one would expect the firm size to affect the extent of technology transfer. Large firms are more able to transfer technology than small firms because large firms possess greater financial and human resources, which may be used for technology transfer. However, large firms with abundant human resources may not have much incentive to transfer technology because parent firms can provide technology to their affiliates whenever technical assistance is needed because of abundant human resources. In other words, large firms can retain control over their affiliates by limiting technology transfer, if they wanted. Based on these observations, the expected sign of the parent firm size (PSIZE), which is measured by the number of workers at parent companies, is ambiguous.

Previous experiences in transferring technology by parent firms should facilitate technology transfer. Indeed, several studies reviewed earlier have confirmed this effect (e.g., Teece 1977, Wakasugi 1996). Because appropriate information is lacking in the METI survey, we use the share of overseas sales in total sales (the sum of sales in Japan and overseas sales) (POSALE) to capture the experiences in overseas activities. POSALE also reflects the extent of internationalization or openness of the parent firms. Considering these factors, we expect POSALE to have a positive impact on technology transfer.

Turning to the characteristics of overseas affiliates, which depend largely on the strategies of their parent firms, especially in the case of Japanese firms, one can think of several factors that could affect the extent of intrafirm transfer of management technology. The length of operation (AAGE) is likely to be an important factor, as was implied by an earlier observation in Table 2. The longer an affiliate has been operating, the greater the extent of technology transfer expected. Local staff at overseas affiliates accumulates experiences over time, which makes it easier for them to absorb technology. Experiences have an important effect on intrafirm technology particularly for Japanese firms, since on-the-job training plays an important role in transferring technology for Japanese firms. Based on these discussions, we expect AAGE to have a positive impact.

The share of equity of an overseas affiliate held (AEQY) by its parent firm has been shown by previous studies to affect the pattern of intrafirm technology transfer, as was discussed in an earlier section. Several studies have shown that the cost of intrafirm technology transfer declines as the share of equity holding by the parent firm increases (see Teece 1977; Ramachandran 1993). The reason behind this relationship is that the threat of the misuse of technologies declines with the level of equity held by parent firms. However, one may expect different relationship between AEQY and the extent of technology transfer. A parent firm may not feel the need to transfer technology to majority-owned affiliates, because the affiliates can depend on their parent company for the supply of technology. There is yet another reason one may posit the negative relationship between equity ownership and the level of transfer of management technology, especially when it is measured by the nationality of the personnel in charge of job assignment. Parent firms tend to assign Japanese staff to responsible positions in a majority owned affiliate because they try to control the management of the affiliate. Furthermore, parent firms assign staff from parent firms, who do not have relevant positions at the parent firms. Such assignment is not unusual for Japanese firms, which practice a life-time employment system. Based on these observations, the expected sign of AEQY is not clear.

Several additional control variables are included in the analysis. The attitude of parent firms and/or their affiliates concerning their relationship with local market in terms of sales and procurement is expected to have impacts on the nationality of personnel in charge of various assignments. One would expect the affiliates with close ties with local market to realize the need to have local staff in responsible positions or the need for technology transfer, in order to have good business relations with local businesses. Because of these reasons we expect the share of local procurement in total procurement of inputs (APROC) and the share of local sales in total sales (ASALE) to have positive relationships with technology transfer.

Finally, we include the size of affiliates (ASIZE) to consider the importance of operation, which is captured by the size in terms of the number of employees, in determining technology transfer. One may conjecture that ASIZE to have a negative impact on technology transfer as parent firms may control the operation of important affiliates by assigning Japanese staff to responsible positions.

As for the variables concerning the host countries, we include two types of variables. One is the level of industrialization (HMFG) and the other is the quality of labor force (HEDU). Both variables represent the capability of labor force to take important positions in the affiliates of Japanese firms. Accordingly, both HMFG, which is measured by the share of manufacturing value added in total value added, and HEDU, which is measured by the attainment ratio for the secondary education, are expected to have positive impacts on technology transfer.

3.2. The Determinants of Intrafirm Technology Transfer: The Results

We conducted regression analyses to test the validity of the arguments presented above concerning the determinants of intrafirm transfer of management technology, which is evaluated by the nationality of staff responsible for the job. If the local staff is in charge, then it is interpreted that technology transfer is achieved. The estimation was conducted for the information obtained from the surveys conducted by the Ministry of Economy, Trade and Industry, Kaigai Jigyo Katsudo Kihon Chosa [Comprehensive Survey of Overseas Activities of Japanese Firms] and Kigyo Katsudo Kihon Chosa [Basic Survey of Business Structure and Activity]. Utilizing the panel dataset constructed from the surveys for 1995, 1998, and 2001, we applied the Random Effect Probit Model, because the dependent variable takes either 0 or 1; 0 for the case where a staff in charge is Japanese, and 1 for the case where a staff is local.

The results of the regression analyses are shown in Table 3, Table 4 and Table 5. The results for all the affiliate in the world are presented in Table 3, those in Asia in Table 4, and those in developed countries (the US and the EU) in Table 5. We analyze the affiliates in Asia and in developed countries separately, because the determinants of transfer of management technology are likely to differ for these affiliates, which in turn are due to the differences in Japanese firms' strategies in these two groups of countries. Specifically, Japanese firms are interested in transferring their management technology to their affiliates in developing countries. By contrast, relatively speaking Japanese firms would be interested in utilizing local management technology at their affiliated in developed countries, since efficient management technology is likely to be available in developed countries. We will examine if such differences exist below.

The results indicate that the characteristics of parent firms, their overseas affiliates and the host countries have important impacts on international intrafirm transfer of management technology. To begin with the characteristics of parent firms, the results show that overseas experiences (POSALE) tend to facilitate transfer of management technology, especially top management, sales, human resource management, and corporate planning.

Table 3. The Determinats of Transfer of Management Technology: Affiliates in the World

Dep variable	CEO	Sales	Purchase	Labor	R&D	Plannnig
# of obs	6502	5651	5560	5877	2062	2762
# of groups	4746	4226	4176	4347	1769	2318
POSALE	0.105	0.437	0.079	0.364	-0.054	1.118
	[0.57]	[2.54]**	[0.47]	[2.32]**	[-0.21]	[4.58]***
PSIZE	-0.052	-0.086	-0.100	-0.049	-0.133	-0.220
	[-2.32]**	[-4.12]***	[-5.00]***	[-2.58]***	[-4.46]***	[-7.51]***
APROC	0.217		0.162	-0.094	0.093	0.051
	[2.53]**		[2.09]**	[-1.29]	[0.79]	[0.46]
ASALE	0.413	0.462		0.475	0.388	0.577
	[4.75]***	[5.82]***		[6.60]***	[3.26]***	[4.98]***
AAGE	0.003	0.001	0.003	0.002	0.017	0.009
	[0.71]	[0.23]	[0.83]	[0.69]	[3.13]***	[1.75]*
AEQY	-0.036	-0.017	-0.017	-0.015	-0.016	-0.017
	[-23.71]***	[-11.66]***	[-12.06]***	[-11.22]***	[-8.16]***	[-8.70]***
ASIZE	0.032	0.061	0.066	0.181	0.038	0.171
	[1.36]	[2.72]***	[3.05]***	[8.63]***	[1.10]	[5.38]***
HMFG	0.015	0.018	0.013	0.016	0.001	0.001
	[2.64]***	[3.61]***	[2.71]***	[3.53]***	[0.15]	[0.10]
HEDU	0.035	0.024	0.022	0.013	0.029	0.019
	[8.00]***	[5.44]***	[5.14]***	[3.24]***	[4.64]***	[3.29]***
C	-6.607	8.160	1.264	7.332	0.660	-0.143
	[-0.00]	[0.02]	[0.98]	[0.00]	[0.71]	[-0.15]

'*','**', '***' indicate statistical significance at 10, 5, 1 percent, respectively.
Random-effect probit model is used for estimation.
Figures in brankets indicate z-ratio.
Industry dummies, year dummies and regional dummies are included.
Source: Authors' estimattion.

Table 4. The Determinats of Transfer of Management Technology: Affiliates in Asia

Dep variable	CEO	Sales	Purchase	Labor	R&D	Plannnig
# of obs	3746	3204	3261	3440	1219	1428
# of groups	2715	2391	2435	2518	1036	1233
POSALE	-0.182	0.089	-0.405	-0.046	-0.041	0.929
	[-0.68]	[0.39]	[-1.83]*	[-0.22]	[-0.12]	[2.84]***
PSIZE	-0.025	-0.067	-0.124	-0.023	-0.156	-0.162
	[-0.82]	[-2.50]**	[-4.86]***	[-0.97]	[-3.97]***	[-4.48]***
APROC	0.442		0.308	0.059	0.196	0.262
	[3.77]***		[3.11]***	[0.63]	[1.24]	[1.83]*
ASALE	0.246	0.157		0.266	0.283	0.460
	[2.04]**	[1.51]		[2.85]***	[1.75]*	[3.11]***
AAGE	0.011	0.008	0.013	0.015	0.031	0.015
	[1.92]*	[1.49]	[2.56]**	[3.09]***	[4.03]***	[2.12]**
AEQY	-0.041	-0.019	-0.018	-0.016	-0.017	-0.017
	[-19.32]***	[-10.56]***	[-10.46]***	[-9.53]***	[-6.60]***	[-7.04]***
ASIZE	-0.090	-0.063	-0.066	0.046	-0.084	0.033
	[-2.56]**	[-2.06]**	[-2.28]**	[1.69]*	[-1.68]*	[0.76]
HMFG	0.021	0.027	0.021	0.023	0.009	0.011
	[3.14]***	[4.74]***	[3.76]***	[4.47]***	[1.03]	[1.48]
HEDU	0.053	0.039	0.031	0.023	0.042	0.038
	[9.36]***	[6.92]***	[5.85]***	[4.57]***	[5.42]***	[5.29]***
C	-6.733	7.942	-7.610	7.170	0.980	-0.341
	[-0.00]	[0.01]	[-0.00]	[0.00]	[0.91]	[-0.34]

'*','**', '***' indicate statistical significance at 10, 5, 1 percent, respectively.
Random-effect probit model is used for estimation.
Figures in brankets indicate z-ratio.
Industry dummies, year dummies and regional dummies are included.
Source: Authors' estimattion.

Large size (PSIZE) is found to discourage transfer of technology for all types of management. As we discussed, large firms have abundant Japanese staff at home, so that they can assign Japanese staff to responsible positions at their overseas affiliates. Large firms tend to assign Japanese staff to important positions at overseas affiliates under their well-developed human resource management system involving not only parent company but also their overseas affiliates. Indeed, many large firms use important positions at overseas affiliates as a proving position for Japanese staff's capability.

Turning to the results on the characteristics of the overseas affiliates, we find that the affiliates with strong local orientation, in terms of their procurement (APROC) and sales (ASALE), to have successfully transferred technology. Between the two, local orientation in terms of sales has a stronger impact on transfer of management technology. These findings are reasonable as one would expect the firms with strong local market orientation to promote local staff for responsible positions.

The size of overseas affiliates (ASIZE) is shown to have a positive impact on technology transfer (Table 3). This finding appears to indicate the strategy of parent firms that assign local staff to responsible positions for the large affiliates. However, this tendency is found only for the affiliates in developed countries (Table 5), and not for those in Asia (Table 4). The contrasting patterns observed for the affiliates in developed countries and those in Asia may reflect the differences in the availability of capable human resources. Specifically, capable employees are available at large affiliates in developed countries, so that local employees are assigned to important positions, while capable employees are in short supply for the affiliates in Asia. Another reason for these contrasting findings for the affiliates in developed countries and those in Asia may be due to difficulty in managing large affiliates in terms of employment by Japanese staff at the affiliates in developed countries because of a variety of requests and demands presented by workers and because of a wide variety of business activities at large affiliates. It is interesting to observe that for the affiliates in Asia the relationship is found positive only for the position in charge of labor management. This finding supports the arguments above in that despite a shortage of capable workers at the affiliates in Asia local staff is given responsibility in labor management at large affiliates because of the difficulty in labor management, which increases with the size of the affiliates. To put it differently, the difficulty in managing workers by Japanese staff arises in at the affiliates in Asia as well as those in developed countries as the size of the affiliate increases.

The length of operation of the affiliates (AAGE) shows a significantly positive impact on transfer of management technology for the affiliates in Asia, but not for the affiliates in developed countries. Indeed, for the affiliates in developed countries AAGE is shown to have a negative impact with statistical significance for the positions of CEO, procurement and labor relations. The finding for the Asian affiliates is consistent with our expectation, and it indicates the importance of experiences in learning management know-how by local staff. The reasons for the finding for the affiliates in developed countries are not clear and thus detailed investigation is warranted. It may be possible that anomalous practices of a few affiliates with very long operation history in developed countries might have caused the result.

The pattern of equity ownership influences technology transfer. The estimated coefficients on AEQY show that a large equity share by Japanese firms discourages transfer of management technology regardless of the location of the affiliates. This observation may be explained by the strategy of Japanese parent firms that for important affiliates, reflected in high equity holding, parent firms assign Japanese staff in responsible positions. This finding is not consistent with the findings from the previous studies on technology transfer, reviewed in an earlier section. One of the reasons for this inconsistency may be due to the differences in the measurement of technology transfer. Earlier studies examined the costs expended for technology transfer, while our study examined the achievement of technology transfer. These contrasting findings may indicate that technology transfer does not necessary depend on the resources spent for technology transfer.

Concerning the factors in the host countries, we found that the level of industrialization (HMFG) has a significantly positive impact on transfer of management technology for the affiliates in the world. This expected result indicates the importance of industrialization experience for assimilating management technology. Quality of labor force (HEDU) is found to have a significantly positive impact on technology transfer only for the affiliates in Asia, and not for the affiliates in developed countries. The finding on the affiliates in Asia is consistent with the finding by Ramachandran (1993), indicating the importance of experiences at the affiliates for technology transfer. The unexpected finding for the affiliates in developed countries may be attributable to small variations in HEDU for developed countries.

In order to examine the robustness of our earlier findings, we conducted a dynamic analysis of the determinants of transfer of management technology for the overseas affiliates of Japanese firms. Taking advantage of the panel dataset, we examined the characteristics of the affiliates that shifted job responsibility from Japanese to local employees. Specifically, we used the information on the affiliates where job responsibility was assigned to Japanese employees in period t (1995 or 1998). Given this dataset we identify affiliates, where job responsibility was shifted to local employees in period t+1, and we assign unity for such affiliates. For the affiliates, where job responsibility remained in the hands of Japanese, we assign zero. This information, either unity or zero, is used as a dependent variable in our regression analysis. The Probit estimation was adopted and the results of the estimation for CEO, sales and procurement tasks are shown in Table 6. One notable finding from the results is the importance of the host country characteristics, namely the level of industrialization and the quality of labor force, for transfer of management technology, particularly for the affiliates in Asia, which is consistent with our earlier findings in Table 4.

Table 5. The Determinats of Transfer of Management Technology: Affiliates in Developed Countries

Dep variable	CEO	Sales	Purchase	Labor	R&D	Plannnig
# of obs	2381	2109	1982	2101	743	1149
# of groups	1754	1582	1505	1580	645	940
POSALE	0.500	0.993	0.513	0.845	-0.591	1.084
	[1.21]	[3.55]***	[1.87]*	[3.08]***	[-1.21]	[2.49]**
PSIZE	-0.145	-0.101	-0.100	-0.108	-0.146	-0.329
	[-2.55]**	[-2.81]***	[-2.84]***	[-2.99]***	[-2.40]**	[-5.02]***
APROC	-0.027		-0.070	-0.291	-0.032	-0.139
	[-0.13]		[-0.52]	[-2.14]**	[-0.14]	[-0.64]
ASALE	0.371	0.628		0.436	0.208	0.338
	[1.82]*	[4.60]***		[3.20]***	[0.89]	[1.54]
AAGE	-0.021	-0.008	-0.014	-0.015	-0.007	-0.001
	[-2.31]**	[-1.43]	[-2.37]**	[-2.70]***	[-0.66]	[-0.13]
AEQY	-0.036	-0.009	-0.011	-0.012	-0.015	-0.016
	[-8.33]***	[-3.33]***	[-3.92]***	[-4.04]***	[-3.52]***	[-3.64]***
ASIZE	0.238	0.186	0.277	0.399	0.241	0.358
	[3.92]***	[4.93]***	[6.82]***	[8.82]***	[3.52]***	[5.19]***
HMFG	0.092	0.039	0.028	0.069	0.084	0.017
	[3.04]***	[1.90]*	[1.37]	[3.22]***	[2.13]**	[0.54]
HEDU	-0.006	-0.017	0.002	-0.011	-0.041	-0.041
	[-0.41]	[-1.78]*	[0.18]	[-1.20]	[-2.05]**	[-2.67]***
C	-12.453	10.227	9.593	-9.569	-9.425	-9.085
	[-6.81]***	[8.27]***	[7.83]***	[-6.83]***	[-5.68]***	[-5.68]***

'*','**', '***' indicate statistical significance at 10, 5, 1 percent, respectively.
Random-effect probit model is used for estimation.
Figures in brankets indicate z-ratio.
Industry dummies, year dummies and regional dummies are included.
Source: Authors' estimattion.

4. CONCLUSIONS

Our analysis of transfer of management technology at overseas affiliates of Japanese firms and their determinants revealed a number of interesting and important observations. As to technology transfer among different job classification, technology has not been transferred much for high-ranking and important positions such as CEOs and deputy CEOs, while technology has been transferred for the jobs requiring close local contact such as labor management.

Statistical analysis of the determinants of transfer of management technology found the importance of the characteristics and strategies of parent firms as well as their affiliates, in addition to the characteristics of the host countries, in determining the extent of technology transfer. Among them, particularly important findings from the perspective of policy makers include the followings. We found that technology transfer is undertaken for the overseas affiliates with close local ties in sales and purchase. The length of operation of the affiliates was found to be important for technology transfer for the affiliates in Asia. Furthermore, we found that technology is transferred for the affiliates located in a host country with high level of industrialization. Availability of high quality of labor is found to promote technology transfer for the countries in Asia.

Several important policy implications can be obtained from our analysis for the countries eager to have management technology transferred. First, the countries have to provide an FDI friendly environment, under which overseas affiliates can operate for long period of time. This is because technology transfer takes time. Second, development of capable workers through education and training is very important, because without them technology transfer is impossible. Third, development of competitive manufacturing sector is important for technology transfer. With competitive local industry, overseas affiliates can increase interaction with local manufacturing firms, which in turn would facilitate not only intrafirm technology transfer, which was the subject of this paper, but also technology spillover to local firms. In order to deal with these policy challenges, the government of developing countries should formulate comprehensive development policies ranging from human resource development to FDI policies and implement them effectively.

Finally, we would like to end this paper by indicating the agenda for future research regarding technology transfer. First, the impacts of technology transfer should be examined. This paper examined the extent of technology transfer achieved but it did not analyze how technology transfer affected the performance of overseas affiliates. Without discerning the impacts of technology transfer on performance such as profitability and productivity, the analysis of technology transfer is not completed. Second, it would be important to conduct an international comparison among MNCs from different countries concerning the extent of technology transfer. Such study would reveal the differences and similarities of technology transfer among them. In this regard MITI (2000) reports that for the affiliates in Asia the proportion of expatriates (staff from the home country) in total CEOs is high at 72 percent for Japanese firms while the corresponding figures for US and German firms are 26 and 23 percent, respectively. These observations are useful but detailed analysis is warranted. Finally, case studies of technology transfer, which would complement our statistical study, should be conducted, in order to deepen our understanding of technology transfer.

Table 6. The Determinatns of the Shift from Japanese to Local Staff: Total Industry

	Affiliates in World			Asia			Developed Countries		
Job	CEO	Sales	Procurement	CEO	Sales	Procurement	CEO	Sales	Procurement
# of obs	1306	595	484	762	387	321	405	155	135
POSALE	0.443	-0.080	-0.347	0.409	0.167	-0.536	0.621	-0.002	-0.255
	[1.21]	[-0.21]	[-0.87]	[0.74]	[0.34]	[-1.03]	[1.00]	[-0.00]	[-0.35]
PSIZE	-0.011	-0.083	-0.053	0.062	-0.102	-0.063	-0.116	-0.033	-0.003
	[-0.28]	[-1.92]*	[-1.11]	[1.11]	[-1.97]***	[-1.07]	[-1.32]	[-0.35]	[-0.03]
APROC	0.054		-0.090	0.253		-0.100	-0.380		-0.320
	[0.32]		[-0.47]	[1.11]		[-0.42]	[-1.00]		[-0.78]
ASALE	0.055	0.385		-0.141	0.155		0.114	0.667	
	[0.33]	[2.23]**		[-0.62]	[0.73]		[0.34]	[1.72]*	
AAGE	0.008	0.005	0.012	0.013	0.003	0.009	0.008	0.019	0.026
	[1.03]	[0.71]	[1.39]	[1.11]	[0.25]	[0.76]	[0.60]	[1.38]	[1.63]
AEQY	-0.014	-0.003	-0.005	-0.018	-0.004	-0.006	-0.005	0.011	-0.001
	[-4.81]***	[-0.81]	[-1.38]	[-4.87]***	[-1.16]	[-1.49]	[-0.68]	[0.98]	[-0.11]
ASIZE	-0.082	0.071	0.070	-0.173	-0.028	-0.085	-0.054	0.150	0.201
	[-1.73]*	[1.49]	[1.33]	[-2.36]***	[-0.43]	[-1.16]	[-0.60]	[1.59]	[2.00]**
HMFG	0.031	0.026	0.024	0.039	0.023	0.022	0.049	0.139	-0.019
	[2.47]**	[2.20]**	[1.86]*	[2.58]***	[1.73]*	[1.53]	[0.66]	[1.25]	[-0.15]
HEDU	0.022	0.024	0.029	0.049	0.024	0.027	-0.029	-0.011	0.027
	[2.42]**	[2.48]**	[2.74]***	[4.20]***	[2.15]**	[2.03]**	[-1.14]	[-0.32]	[0.78]
C	-0.715	-0.954	-1.325	-1.069	-0.042	-0.459	0.071	-5.002	-1.309
	[-0.84]	[-1.08]	[-1.42]	[-1.07]	[-0.04]	[-0.42]	[0.04]	[-2.03]***	[-0.43]

Appendix Table 1. Basic Information on the Varibles used in the Analysis

Categories	Variable Name	Explanation/Definition	Source	N	mean	sd	p25	p75
Overseas Affiliates	TT-CEO	Chief Executive Officer	METI, Kaigai Jigyo Kihon Chosa	6502	0.211	0.408	0	0
	TT-Deputy	Deputy CEO	METI, Kaigai Jigyo Kihon Chosa	5603	0.403	0.491	0	1
	TT-LAB	Labor management	METI, Kaigai Jigyo Kihon Chosa	5877	0.683	0.465	0	1
	TT-Acct	Accounting	METI, Kaigai Jigyo Kihon Chosa	6115	0.577	0.494	0	1
	TT-Sales	Sales	METI, Kaigai Jigyo Kihon Chosa	5651	0.550	0.498	0	1
	TT-Proc	Procurement	METI, Kaigai Jigyo Kihon Chosa	5560	0.625	0.484	0	1
	TT-R&D	Research and development	METI, Kaigai Jigyo Kihon Chosa	2062	0.538	0.499	0	1
	TT-Plan	Planning	METI, Kaigai Jigyo Kihon Chosa	2762	0.572	0.495	0	1
	APROC	Local procurement/total procurement	METI, Kaigai Jigyo Kihon Chosa	6502	0.449	0.390	0.037	0.871
	ASALE	Local sales/total sales	METI, Kaigai Jigyo Kihon Chosa	6502	0.651	0.399	0.229	1
	AEQY	Equity owned by Jap firms/total equity	METI, Kaigai Jigyo Kihon Chosa	6502	0.835	0.240	0.65	1
	AAGE	Length of operation	METI, Kaigai Jigyo Kihon Chosa	6502	11.523	8.890	5	16
	ASIZE	Ln (number of employees)	METI, Kaigai Jigyo Kihon Chosa	6502	4.345	1.749	3.135	5.617
Parent companies	POSALE	Overseas sales/Total sales	METI, Kaigai Jigyo Kihon Chosa	6502	0.296	0.186	0.136	0.440
	PSIZE	Number of employees	METI, Kigyo Katsudo Kihon Chosa	6502	7.763	1.565	6.652	8.840
Host countries	HMFG	Manufacturing VA/GDP	World Bank, World Development Indicators CD-ROM	6502	0.234	0.079	0.176	0.309
	HEDU	Secondary education attainment	Barro & Lee	6502	17.044	7.541	12.35	22.28

Note: The value for technology transfer (TT) is unity, if the staff in charge of assignment is local. Otherwise, zero.

Appendix Table 2. Correlation Matrix of the Variables

	TT-CEO	TT-Deputy	TT-LR	TT-Acct	TT-Sales	TT-Purch	TT-R&D	TT-Plan	APROC	ASALE	AEQY	AAGE	ASIZE	POSALE	PSIZE	HMFG	HEDU
TT-CEO	1																
TT-Deputy	0.0109	1															
TT-LR	0.3038	0.2825	1														
TT-Acct	0.3273	0.2703	0.589	1													
TT-Sales	0.3658	0.293	0.4521	0.4842	1												
TT-Proc	0.2928	0.2535	0.5584	0.5505	0.6177	1											
TT-R&D	0.3539	0.2553	0.4337	0.4425	0.4916	0.5475	1										
TT-Plan	0.3478	0.281	0.4687	0.4669	0.5541	0.5835	0.682	1									
APCH	0.1708	0.0288	0.0358	0.0748	0.0842	0.1081	0.057	0.0794	1								
ASALE	0.1319	0.0613	0.1549	0.1017	0.1522	0.1312	0.0954	0.0937	0.0818	1							
AEQY	-0.5133	-0.104	-0.2102	-0.2387	-0.2161	-0.1708	-0.1764	-0.1794	-0.2337	-0.1475	1						
AAGE	-0.0103	0.0015	0.0028	-0.0361	0.0364	0.015	0.0452	0.0445	-0.0391	0.0985	0.0776	1					
ASIZE	0.0433	-0.0546	0.1303	-0.0745	0.0126	0.036	-0.022	0.0211	0.0696	-0.0776	-0.1123	0.178	1				
POSALE	-0.0202	0.0002	-0.0072	-0.1042	-0.0533	-0.0739	-0.0306	-0.0116	-0.1282	0.0987	0.0139	0.1282	0.2217	1			
PSIZE	0.0394	-0.0352	-0.0068	-0.2029	-0.0268	-0.0802	-0.0667	-0.0901	-0.0928	0.1122	-0.0628	0.1306	0.2806	0.3395	1		
HMFG	0.1163	0.0312	0.0746	0.0698	0.0226	0.008	-0.0124	0.023	0.0758	-0.1659	-0.3305	-0.223	0.2312	-0.0838	-0.0851	1	
HEDU	0.0417	0.0728	0.0098	0.0223	0.043	0.0818	0.0959	0.0418	-0.0051	0.1067	0.1328	0.0744	-0.1727	-0.0051	-0.0488	-0.2823	1

Source: Authors' calculation.

ACKNOWLEDGEMENTS

The authors thank helpful comments and discussions from the participants on the Workshop on Multinational Corporations: Their Behavior and Economic Impacts at RIETI, December 10, 2005 and seminar at RIETI, January 24, 2006. The authors are particularly indebted to Drs. Eric Ramstetter and Rene Belderbos and Mr. Yuji Hosoya for their very helpful comments.

REFERENCES

Beldeerbos, Rene A. and Marielle G. Heijltjes (2005) "The Determinants of Expatriate Staffing by Japanese Multinationals in Asia: Control, Learning, and Vertical Business Groups," *Journal of International Business Studies*, 36 (3), pp.341-354.

Davies, Howard (1977) "Technology Transfer through Commercial Transactions, *Journal of Industrial Economics* 26 (December), pp. 161-75.

Javorcik, Beata Smarzynska and Mariana Spatareanu (2005) "Disentangling FDI Spillover Effects: What Do Firms Perceptions Tell Us?" in T.H.Moran, E.M.Graham and M Blomstrom eds. *Does Foreign Direct Investment Promote Development?* Institute for International Economics, Washington, D.C, pp. 45-71.

Ministry of International Trade and Industry (2000) *White Paper 2000* (in Japanese).

Ramachandran, Vijaya (1993) "Technology Transfer, Firm Ownership, and Investment in Human Capital," *Review of Economics and Statistics* 75 (November), pp. 664-70.

Saggi, Kamal (1999) "Trade, Foreign Direct Investment, and International Technology Transfer: A Survey," Policy Research Working Paper Series 2349, the World Bank.

Teece , David J. (1977) "Technology Transfer by Multinational Firms: The Resource Cost of Transferring Technological Know-how, *Economic Journal* 87 (June), pp. 242-61.

Urata, Shujiro (1999) "Intrafirm Technology Transfer by Japanese Multinationals in Asia," in Dennis J. Encarnation ed.. *Japanese Multinationals in Asia: Regional Operations in Comparative Perspective*, Oxford University Press, New York, 1999

Wakasugi, Ryuhei (1996) "Gijutsu Iten no Kettei Yoin [The Determinants of Technology Transfer], in *Kagai Chokusetsu Toshi to Nihon Keizai [Foreign Direct Investment and the Japanese Economy]*, ed. Sueo Sekiguchi and Hiroshi Tanaka, Toyo Keizai Shimpo Sha, pp. 98-110.

Yamashita, Shoichi (1991)'Economic Development of the ASEAN Countries and the Role of Japanese Direct Investment,' in Sohichi Yamashita ed. *Transfer of Japanese Technology and Management to the ASEAN Countries*, University of Tokyo Press, pp.3-22.

In: Micro-Evidence for the Dynamics of Industrial Evolution ISBN: 978-1-60456-388-7
Editors: A. Heshmati, J.-D. Lee

Chapter 7

THE RELATIONSHIP BETWEEN INNOVATION AND THE PERFORMANCE OF KOREAN MANUFACTURING FIRMS

Inha Oh*

Technology Management, Economics and Policy Program,
College of Engineering, Seoul National University, Seoul, Korea

Sangyun Han†

I&TB, CUNO, Marketing, 3M Korea Co. Ltd., Seoul, Korea

Almas Heshmati‡

University of Kurdistan Hawler, and
Hawler Institute for Economic and Policy Research
Hawler, The Federal Region of Kurdistan, Iraq

ABSTRACT

The manufacturing industry has been a principal contributor to growth in the Korean economy for forty years. Information technology related manufacturing sectors and conventional heavy industries in particular are still major sources of current economic growth. Technology innovation has been recognized as a critical factor in the productivity growth of firms. Therefore, the Korean government is using its capacity to induce innovation in manufacturing industry and through various support and subsidy policies, it aims to improve the R&D intensity of firms to enhance productivity growth in manufacturing. However, the actual proof of a positive relationship between R&D inputs and productivity growth in the manufacturing sectors is not obvious. This has resulted in difficulties designing public subsidy policies. This paper tries to investigate and analyze the relationship between innovation and the performance of Korean manufacturing firms.

* Corresponding author: Inha Oh, whistle3@snu.ac.kr, inha.oh@gmail.com
† E-mail: lucides9@snu.ac.kr
‡ E-mail: almas.heshmati@ukh.ac

In particular, this paper investigates the links and causality directions between innovation input and output, and also between the innovation output and labor productivity growth at the firm level. In addition, this paper investigates which environmental factors have an effect on the innovation itself. Recent improvements in econometric methodologies, dealing with simultaneity and selection bias, are also taken into account. The empirical result is based on the Korean Innovation Survey 2002 (KIS2002) combined with financial data from firms. Results suggest that the actual linkage going from R&D investment to the innovation output was not observed. Policy makers must reflect on this issue before designing policy measures to enhance productivity.

Keywords: Innovation Input, Innovation output, Productivity, Korean Manufacturing
JEL Classification Numbers: C33, E22, L60, O32, N65

1. INTRODUCTION

The manufacturing industry of Korea has played a central role in the development program of the Korean government. In the first quarter of 2005, the share of the manufacturing industry in gross value added was 32 %, while it was only 7.6 % in 1970. Since the 1980s, more than 90% of exports from Korea have originated from its manufacturing industry. Thus, information on the relationship between innovation and the performance of Korean manufacturing firms was essential for designing the industrial policy of Korea.

The technology innovation of manufacturing industry and productivity growth are important factors for the long-term survival and growth of firms. And as a proxy variable for the technology innovation, R&D activity was known to be the source for the heterogeneity in the firm productivities. Cohen and Levinthal (1989) divided the purposes of R&D activities as a technology innovation and accumulation of experience by learning. They defined innovation as the creation of novel knowledge and information. The innovation at the firm level involves the creation of new products or the introduction of novel processes to the market, which will eventually enhance firm performances and productivity. In this regard, European countries have utilized the standardized community innovation survey[1] (CIS) to observe the degree of technology innovation, productivity and firm performances and their relationships (for examples, Evangelista et al., 1997; Lööf and Heshmati, 2002 and 2006). Particularly, the level of labor productivity of Korean manufacturing industries in year 2004 was known to fall behind by 1.5 to 2 times compared to USA manufacturing industries, according to the report from Korea Productivity Center (2004). For that reason, the Korean government has tried to stimulate R&D with the objective of inducing technology innovation and productivity growth.

Based on the Korean innovation survey[2] in the year 2002 (henceforth, KIS2002), Shin et al. (2002) reported that the top three main obstacles, which Korean manufacturing firms encountered during the process of technology innovation, were the lack of developed human resources, the difficulty of accessing technology information and the lack of potential to

[1] The CIS means the group of countries which voluntarily participated in the common innovation surveys based on the Oslo manual, as well as the questionnaires of the survey or the survey activity itself.

create technology innovation. Meanwhile, competition among domestic firms, difficulties in handling risks and uncertainties and long term investment recovery were seen as obstacles and external factors to the firms' innovation environment. When we look at the situation by size of firms, the large-sized firms answered that lack of ability to manage R&D process was the main obstacle, while the small and medium sized firms named the risks of technology imitation by local competitors and uncertainties aroused from lack of the technology information as the main obstacles. It seems that the innovative environment of the industry and its ability in management and planning are also very important factors in fostering innovation in the industry other than with the provision of R&D funds. However, according to the webpage of the Korea Industrial Technology Association (http://www.koita.or.kr), the lists of policies aimed at fostering R&D activities (and subsequent technology innovations) are mainly concentrated on the financial support to promote the R&D expenditure of supported firms, other than by enhancing the innovative environment.

With reference to the above issues, this research is aimed at observing whether the inputs to facilitate innovation, i.e. R&D investment, really increase the performance of firms (innovation output) and further labor productivity growth in the Korean manufacturing industry. The dataset applied here is KIS2002 combined with the financial statements of firms obtained from the database of the Korea Information Service (henceforth, KISinfo). This enabled us to increase the reliability of the database and use more detailed financial terms of firms under analysis. The results suggest that the policies introduced to stimulate technology innovation should be separated from conventional support aimed at the R&D expenditure of firms. The former should be introduced such that they increase the productivity of firms.

The rest of the paper is organized as follows. The literature is reviewed in the next section. The theoretical model with a focus on the three interrelated equations with correction for selectivity and simultaneity biases is outlined in Section 3. Section 4 describes the KIS2002 data and variable definitions. Section 5 discusses the empirical model used in practice. In Section 6, the empirical results are discussed. The final Section 7 concludes this study.

2. Review of the Literature

The previous literature dealing with the relationships between R&D investment, innovation and productivity were mostly focused on the linkage between R&D investment and one of the firms' performances or productivity growth. They mainly observed the effect of R&D investment when regressed on a firm's performance or productivity growth, controlling for firm characteristics such as age, industrial sectors, patents, sales and many others. Accordingly, Arrow (1962) and Solow (1997) showed that the differences of R&D investments of firms were the main source of firm heterogeneity in productivity growth, and emphasized the role of the process of learning by doing in technology innovation. Cohen and Levinthal (1989) also pointed out the importance of R&D investment to the technology innovation of firms. They argued that R&D activity creates novel knowledge and information which will induce technology innovation which results in an introduction of novel products or

[2] It was a counterpart of Oslo manual in Korea conducted targeting the manufacturing industries.

processes. It will also induce the accumulation of knowledge capital inside firms and improve firms' performances.

Cohen and Klepper (1996) econometrically showed that R&D activity had direct effects on the growth of the productivity and size of firms. Baldwin and Hanel (2003) also utilized the Canadian innovation survey to show that 63% of firms successful in technology innovation experienced growth in profitability. They found the sources of profitability improvement from cost reduction and the introduction of novel products to the market due to the R&D activities.

When looking at the technology innovation in the framework of endogenous growth model, Aghion and Howitt (1998) developed a Schumpeterian version of the endogenous growth model. Their model predicted that long-run growth should be positively correlated with R&D, productivity, flow of patents and new products and it should decrease with the rate of depreciation of human and physical capital. Inspired by the macro models of endogenous growth, in particular the quality ladder model, Klette and Griliches (1997) proposed a model of endogenous firm growth. In their model R&D and innovations were the engines of growth. The idea was promising as a benchmark model to understand the established stylized facts of firm behavior. The main propositions derived from the model were consistent with many findings from empirical studies: R&D increases proportionately with firm size; R&D intensity is positively related to the profit per unit of output; R&D follows a random walk (Gibrat's law in R&D); the model will generate a highly skewed distribution of firm size; and industries with higher innovative opportunities, greater willingness to pay for higher quality products and larger demand will have higher R&D intensities.

The previous literature related to the R&D activities as shown considered R&D investment as an input factor of production and observed how the level of the R&D investment of firms affected productivity and the performance of firms, and how the effects were different by size or the industry which the firm belonged to. However, other research concerned the characteristics of R&D activities. Mairesse and Sassenou (1991), in their study with firm level data, concluded that it was hard to find a statistically significant relationship between R&D investment and productivity growth. They found the reason from the time lags between the point of time that investment occurred and the realization of the effects as productivity growth. They also pointed out that the effect of learning by doing, accumulated along with production experience tended to increase productivity, which made it hard to separate the independent effect of R&D activity. Hall and Mairesse (1995) also mentioned the time lags between the investment and its realization. However, based on the firm level data of the French manufacturing industries, they found that firms with past R&D experiences showed higher elasticity of R&D on firm performance. The firms which accumulated knowledge and information through R&D experiences seemed to enjoy higher returns on R&D investments.

As shown before, to observe the effect of R&D activity on the future growth of productivity or the performance of firms, there exist lots of barriers to overcome apart from the issue of measurement error. Perceiving R&D investment as an innovation input, and focusing the empirical analyses on the relation of innovation input with innovation output, which results from technology innovation, and innovative environment surrounding firms, many studies have failed to see technology innovation as a simple input/output process and tended to consider other environmental variables as well as R&D investment.

Evangelista et al. (1997), by using the Italian innovation survey, analyzed the degree and characteristics of innovation by industrial sectors and firm sizes. Through the research, they found that R&D activity is one of the decisive factors in technology innovation and an important intangible asset for innovation. However, only one third of investments in R&D were directly affected in a way which led to innovation and the remaining factors indirectly affected the innovation process through the innovation form as embodied technologies. Later Crépon et al. (1998) suggested the new econometrics framework to analyze the relationship among R&D, innovation and productivity, based on the Cobb-Douglas production function. Lööf and Heshmati, (2002 and 2006) developed the model further. It was based on the interrelated equations consisting of innovation input, innovation output and productivity equations. The main contributions of these studies could be found in handling the problem of selectivity and simultaneity biases.

This paper directly adopts the methodology suggested by Lööf and Heshmati, (2002 and 2006) to avoid possible biases and utilizes a great deal of information gained from the Korean innovation survey. The financial statements of firms were also obtained from other reliable data sources[3].

3. The Theoretical Model

The empirical model adopted in this research originated from Crépon et al. (1998). They considered innovation output, which was products that resulted from innovation, as a source for productivity growth. Hence, their model includes three relationships; (i) innovation inputs and its decision factors, (ii) the knowledge production function which relates innovation input and innovation output, and (iii) a function which relates productivity growth and innovation output. Lööf and Heshmati (2002 and 2006) further expanded the model by Crépon et al. (1998) to allow for a simpler correction of selectivity biases and simultaneity problems.

The model was made up of four equations. The first and second equations are as follows:

$$\begin{aligned} &g_i = x_i^0 \beta_0 + u_i^0, \ \ g_i = 1 \text{ if } g_i > 0, \text{else } g_i = 0, \\ &k_i \big| g_i > 0 = x_i^1 \beta_1 + u_i^1, \ k_i = k_i \text{ if } k_i > 0, \text{else } k_i = 0, \end{aligned} \tag{1}$$

where, g_i is a dummy variable indicating the existence of R&D expenditure of each firm, k_i is the actual level of R&D expenditure of firm *i*. The x_i^0 and x_i^1 are observed characteristics of firm *i*. The β_0 and β_1 are vectors of unknown parameters to be estimated reflecting the impact of certain factors on the probability of engagement in R&D activity and the actual level of R&D expenditure. The u_i^0 and u_i^1 are random error terms with mean zero, constant variances and are not correlated with the explanatory variables. However, the two error terms are correlated with each other. The correlation coefficient suggests a selection based on the unobservable characteristics of firms with a high potential to be involved in R&D activities.

[3] It was brought from the Korea Information Service (KISinfo). All the financial information of externally audited Korean firms was accumulated in the KISinfo database.

The model (1) can be estimated in a two-step sample selection procedure. The first step is probit and the second step is ordinary least squares method, i.e. Heckman's two step procedure. The inverted Mill's ratio obtained from the first equation will be included as explanatory variables in the second step with only observations with positive R&D investments.

The second model, as described in Griliches (1994), observes the transformation process that goes from innovation input to economically valuable innovation output as the knowledge production function written as:

$$t_i^* = \alpha_k k_i^* + x_i^2 \beta_2 + u_i^2, \tag{2}$$

where t_i^* is the observed innovation output (sales), and α_k is the elasticity of innovation output with respect to innovation input. x_i^2 is a vector of explanatory variables including Mill's ratio from the probit equation to correct for a possible selection bias, and u_i^2 is the random error term with means zero and constant variance and is not correlated with the x_i^2 variables.

The last equation in the model describing the relationship between innovation output and productivity growth is labeled as the productivity equation expressed as:

$$\dot{q}_i = \alpha_t t_i^* + x_i^3 \beta_3 + u_i^3, \tag{3}$$

where $\dot{q}_i$ is the rate of productivity growth, α_t is the elasticity of productivity with respect to innovation output, and x_i^3 is a vector of other common explanatory variables in a Cobb-Douglas productivity function framework.

4. Data and Variables

4.1. The Data

This study is based on the KIS2002, which is the counterpart of the innovation study conducted following OECD Oslo manuals for Korean manufacturing industry in 2002. The KIS2002 was obtained from randomly distributed questionnaires following Neyman optimum allocation, considering the industrial sectors and number of employees. Among all firms which received the surveys, 60.6% answered the questionnaires which resulted in 3776 respondent firms. These observations are connected to the externally audited firm database of KISinfo, based on the firm registration numbers. We dropped observations with missing data in the major financial statements as sales and exports. We also eliminated observations with missing data in the innovation survey, i.e. if the firm did not answer whether they had results from innovation, the source of innovation and others. Finally 1030 observations of firms were left for analysis. The database represents Korean manufacturing firms receiving an external audit.

4.2. The Variables

First, we will explain the dependent variables used in this research. The innovation input was defined as the spending on R&D. For the first probit model, the dependent variable was given the value of zero if the R&D expenditure was not positive and the value of one if the R&D expenditure was positive. The Mill's ratio was calculated as the ratio of density and distribution function based on this probit regression. The second OLS regression was conducted using the level of log of R&D expenditure per employee as the dependent variables. The predicted level of R&D expenditure calculated based on the second OLS regressions was used in the other two interrelated equations which are the knowledge production function and productivity equation. The dependent variable for knowledge production function is the innovation output, and it was substituted by the amount of innovation output, which means sales originated from technology innovation. The productivity in the last production function was calculated as the log of value added per employee, which was value added labor productivity.

The independent variables used in this research are defined as follows. The number of patents pending drawn from KIS2002, contains the number of applied patents during all processes of innovation in the period from 2000 to 2001. The level of foreign equity ownership, drawn from KISinfo, was defined as average value in 2001. The levels of export, wage and tangible assets are calculated per employee based on year 2001.

The questionnaire of KIS2002 provides various firm characteristics related to the firm innovation environment. However since there were 86 questions, all these variables cannot enter the model. Following Heshmati (2006) we grouped the questions into seven main categories and composite indices were computed using factor analysis with varimax rotation. The questionnaire was constructed such that high values of the indices have positive relationship to the performance of firms. The categories are: firm strategies (15 questions), internal sources of knowledge (7 questions), external sources of knowledge (20 questions), location of cooperation (13 questions), importance of cooperation (9 questions), hampering project innovation factors (14 questions), and hampering innovation factors (8 questions). The first eigenvalues always exceeded 5.3. Hence, the first factors were used to make composite indices and they explained more than 56% of the total variance at the minimum case. All indices were normalized to have mean zero and variance one, and were used as a control for the innovation environment of each firm.

Merger, closing, transfer and venture are dummy variables used in the specification of the model of Korean firms. Venture indicates whether firms are designated as a venture company or not. The variable transfer indicates whether firms have experienced the technology transfer from other companies or institutions during 2000 and 2001. Merger and closing indicate whether mergers and closings have increased or decreased the firm's growth in sales by more than 10% per annum comparing the years 1999 and 2001.The dummy variable, indicating process innovation, was also added to distinguish process innovation from product innovation.

Lastly, three groups of dummy variables are used. Three size dummies, nine industry dummies and five regional dummies are used in order to capture the firm and industry heterogeneities. Appendix 1 shows the definition and calculation of variables used for the dependent and independent variables in the model in more detail[4].

Table 1 shows descriptive statistics of variables used, where we compare firms engaged in R&D and those not. Firms doing R&D showed higher values in innovation output as expected and larger firms tended to engage more often in R&D. The innovation potential is never unrelated to the size of firms, and technology oriented venture businesses tend to be small at the entry stage to the market. Small firms, as they increase their size and accumulate experience, can stimulate exports and innovations. As pointed out in Arrow (1962) and Solow (1997), past experiences would enhance firm performance through learning by doing. This effect can also be found in Klette and Griliches (1997). Firms with higher foreign equity ownership were doing more R&D. There existed contradictory arguments about the R&D intensity and foreign ownership. Ranga Chand (1982) analyzed the Canadian manufacturing industries and concluded that Canadian controlled domestic corporations are more R&D intensive than their foreign-controlled counterparts, while Zhang et al. (2003) showed in their analyses on Chinese firms, that foreign firms have higher R&D and productive efficiency than domestic collective-owned enterprises and joint stock companies. The differences seemed to originate from the different circumstances of each country. The number of patents pending, which was regarded as an output of innovation in much research (Griliches, 1990; Crépon et al., 1998) also showed significant differences between firms doing R&D and non-R&D firms. Firms with a venture business certificate showed a high level of R&D expenditure, since the certification itself requires a certain level of R&D investment. Most of the composite indices showed significant differences between firms doing R&D and those which were not, which means the innovation environment also affected significantly the R&D investment decision of firms. Firms located in different industrial sectors and regions showed heterogeneity in R&D conductions.

5. THE EMPIRICAL MODEL

The econometric model used in this research is based on Crépon et al. (1998) and Lööf and Heshmati (2002 and 2006). The system consists of four equations. The first two equations represent innovativeness and they are estimated by probit with both innovative and non-innovative firms and least squares with only innovative firms, respectively. The interrelated third and fourth equations are estimated as a system using the three-stage least squares (3SLS) method. These equations are estimated based on observations with a strictly positive innovation input. The four equation model is written as:

[4] The labels for the variables defined in Appendix 1 were used for further writing.

Table 1. Descriptive Statistics of R&D firms and Non-R&D firms

Variables	All firms MEAN	All firms STD	R&D firms MEAN	R&D firms STD	Non-R&D firms MEAN	Non-R&D firms STD	t-test
INNOVIN	2912.67	7187.40	4702.27	8661.43	0.00	0.00	***
INNOVOUT	0.31	0.36	0.36	0.37	0.22	0.33	***
VADD	61467.89	64179.74	60851.88	48131.84	62470.48	84049.28	
SALES	282681.24	328532.39	277946.31	346358.57	290387.58	297512.23	
SIZE	279.10	940.36	350.20	1177.60	163.36	215.01	***
EXPORT	17366.27	62165.89	19793.38	63234.63	13416.03	60256.83	
PROFIT	50614.39	63282.02	50262.57	47590.56	51187.00	82756.58	
WAGE	10853.50	9888.79	10589.31	7640.18	11283.48	12726.09	
CAPITAL	279938.30	367957.46	273148.16	373781.00	290989.59	358474.66	
FEQUITY	0.81	4.09	1.20	4.89	0.17	2.07	***
PATENT	3.01	17.09	4.26	21.23	0.96	5.23	***
MERGER	0.06	0.23	0.07	0.25	0.04	0.20	
CLOSING	0.03	0.18	0.03	0.17	0.04	0.19	
VENTURE	0.26	0.44	0.35	0.48	0.11	0.31	***
TRANSFER	0.28	0.45	0.34	0.47	0.18	0.39	***
PROCINNOV	0.43	0.50	0.49	0.50	0.35	0.48	***
CFS	0.00	1.00	0.13	0.98	-0.21	0.99	***
CISK	0.00	1.00	0.07	1.01	-0.12	0.97	***
CESK	0.00	1.00	0.21	1.04	-0.34	0.82	***
CLCO	0.00	1.00	0.06	1.05	-0.10	0.90	**
CICO	0.00	1.00	0.13	1.09	-0.21	0.79	***
CHPF	0.00	1.00	0.11	0.98	-0.19	1.00	***
CHIF	0.00	1.00	0.14	1.07	-0.22	0.82	***
IND1	0.04	0.21	0.04	0.19	0.06	0.23	
IND2	0.06	0.25	0.05	0.22	0.08	0.27	*
IND3	0.04	0.19	0.01	0.12	0.08	0.27	***
IND4	0.21	0.41	0.23	0.42	0.18	0.38	**
IND5	0.20	0.40	0.17	0.37	0.26	0.44	***
IND6	0.09	0.29	0.10	0.30	0.08	0.27	
IND7	0.20	0.40	0.25	0.43	0.14	0.35	***
IND8	0.12	0.33	0.13	0.33	0.12	0.32	
IND9	0.02	0.14	0.02	0.14	0.02	0.14	
AREA1	0.53	0.50	0.55	0.50	0.49	0.50	**
AREA2	0.11	0.32	0.11	0.31	0.11	0.32	
AREA3	0.05	0.23	0.06	0.24	0.05	0.21	
AREA4	0.29	0.46	0.27	0.44	0.33	0.47	**
AREA5	0.01	0.12	0.01	0.09	0.02	0.15	
SIZE1	0.33	0.47	0.27	0.45	0.41	0.49	***
SIZE2	0.34	0.47	0.33	0.47	0.35	0.48	
SIZE3	0.33	0.47	0.40	0.49	0.23	0.42	***
Number of Observation	1030		638		392		

***, **, and * indicates 1%, 5%, and 10% significance level, respectively. T-test was conducted comparing R&D firms and Non-R&D firms considering equity of variance. The variables in this table did not take log values. Monetary unit is in thousand Korean Won.

$$INNOV_i = \beta^1{}_0 + \sum_n \beta^1{}_n X^1{}_{ni} + \varepsilon^1{}_i \quad (4)$$

$$INNOVIN_i = \beta^2{}_0 + \sum_m \beta^2{}_m X^2{}_{mi} + \beta_{MR} \hat{MR}_i + \varepsilon^2{}_i \quad (5)$$

$$INNOVOUT_i = \beta^3{}_0 + \sum_l \beta^3{}_l X^3{}_{li} + \beta_{MR} \hat{MR}_i + \beta_{INNOVIN} INNO\hat{V}IN_i + \beta_{PROD} \hat{\dot{P}}ROD_i + \varepsilon^3{}_i \quad (6)$$

$$\dot{P}ROD_i = \beta^4{}_0 + \sum_j \beta^4{}_j X^4{}_{ji} + \beta_{INNOVOUT} INNOV\hat{O}UT_i + \beta_{PROD^{-1}} PROD^{-1}{}_i + \varepsilon^4{}_i, \quad (7)$$

where the variables marked with a hat represent predicted values, MR is an inverted Mill's ratio introduced to correct for possible selection bias, $PROD^{-1}$ is a lagged labor productivity (year 2000), $\dot{P}ROD$ is a growth in labor productivity comparing 2000 and 2001[5], X are explanatory variables including various firm characteristics, composite indices and dummy variables.

β^1 and β^2 are vectors of unknown parameters to be estimated reflecting the impact of certain factors on the probability of being engaged in R&D and other innovation investments and on the actual level of these investments. The β^3 are estimated parameters associated with the level of innovation output while β^4 are associated with the determinants of labor productivity growth.

For estimating equation (4) and (5), in order to correct for the sample bias problem, Heckman's two-step estimation procedure can be applied. In the first stage, a probit regression is computed in order to estimate the probability of engaging in R&D activity. This regression is used to estimate the inverted Mill's ratio for each firm, which is used as an instrument in the second regression. From Heckman's two-step estimation procedure, $\hat{MR}$ and $INNO\hat{V}IN$, are obtained and used as explanatory variables in equation (6). Equations (6) and (7) are then estimated based on a sample of innovative firms with positive R&D expenditure. One problem here is that some of the explanatory variables are often determined jointly with the dependent ones. For example, the innovation input is endogenous in the innovation output equation (6), and innovation output is endogenous in labor productivity equation (7). In order to derive a consistent estimator, we account for simultaneity by relying on the instrumental variable approach. The instruments consist of variables not correlated with the model error terms but correlated with the endogenous variables. The instruments are described in Section 4.1 as explanatory variables of equations (6) and (7) and their lagged values, in different combinations.

[5] Some observations had missing values for the wage level of firms in 2000. So they were roughly estimated assuming the share of total wage in sales in each firm persisted throughout years 2000 and 2001.

It should be noted that, the above approach introduced by Lööf and Heshmati (2002) in addition to the simplification of the estimation procedure, splitting the four equations into two parts, not only allows for within-part correlation but also for limited between-part correlation among the error terms facilitated through the inclusion of $\hat{MR}$ in the second part. This approach is thus an intermediate approach compared to the Pakes and Griliches (1984) model which neglects correlations and the Crépon et al. (1998) approach allowing for full correlation among the four equations.

5.1. Specification of the Model

In using cross-sectional data in the context of innovation, we observe R&D investment only for a single year. We must therefore assume that the level of investments in one year can be used as a proxy for long-term R&D investment. This presumes that firms do not experience major fluctuations in their R&D investment behavior. There is evidence that R&D expenditure is highly correlated from one year to another (Griliches, 1988).

The determinants of innovation input labeled in the equation (4) consist of *SIZE, SALES, EXPORT, WAGE, FEQUITY, MERGER, CLOSING, VENTURE, TRANSFER,* composite indices and size, region and industry dummy variables, which are summarized in Appendix 1.

The explanatory variables in the selection equation (5) consist of *SIZE, SALES, EXPORT, WAGE, EXPORT, CAPITAL, PATENT, MERGER, CLOSING, VENTURE, TRANSFER,* estimated inverted Mill's ratio, composite indices and size, region and industry dummy variables.

The determinants of innovation output in equation (6) consist of the predicted value of *INNOVIN*, the predicted value of *PROD* growth, *SALES, FEQUITY,* estimated inverted Mill's ratio, *MERGER, CLOSING, VENTURE, TRANSFER, PATENT, PROCINNOV*, composite indices and size, region and industry dummy variables.

The explanatory variables entering the labor productivity equation (7) contains information on predicted value of *INNOVOUT*, lagged labor productivity, *SIZE, CAPITAL, EXPORT, MERGER, CLOSING, VENTURE, TRANSFER, PATENT,* composite indices and size, region and industry dummy variables.

6. The Empirical Results

The empirical results in this study are based on the KIS2002 and KISinfo datasets described earlier. The resulting data is cross-sectional data covering year 2001 (also 2000 for some innovation indicators). The total number of observations used in the estimation steps is 1030. For the specification of the innovative model estimated by the Heckman two-step procedure, availability of data and the relationship between explanatory variables and their expected relationship to the decision of innovation investment as well as individual variables the significance level determined the final model specification. The remaining models of innovation output and productivity growth are estimated jointly as a system using the 3SLS estimation method. The sample is restricted to firms doing R&D, but it accounts for possible selection effects. Many of the innovation activities, obstacles and cooperation strategy indicators were formulated as complex questions in the questionnaires. In order to avoid a large number of dummy variables we estimated composite indices and they were used in the entire stage of the model specification as controls variables for innovation and performance of firms.

6.1. The Innovation Input Results

Table 2 shows the results from the Heckman's two step estimation of innovation input model. First we look at the table for the decision to engage in R&D (probit column). The level of wages per employee is negatively associated with the probability of engagement in R&D activities. Also, firms in wooden products, pulp, paper, publication and press industries are less likely to engage in R&D activities (where food and tobacco industries are used as a base). However the probability of engaging in R&D increases with firm size measured in terms of sales and the number of employees as stated in Cohen and Klepper (1996), which argued that the likelihood of a firm reporting positive R&D effort rises with firm size and approaches one for firms in the largest size ranges. Firms with a higher ratio of foreign equity tend to engage in R&D more often, and firms with a venture business certificate and firms with technology transferred from other institutions also show a higher probability of engaging in R&D.

Now we turn to the innovation input (or R&D investment per employee) equation (OLS column). In the second step of the model, the estimation of innovation input is based only on firms doing R&D which have a different probability of innovative activities compared to non-innovative firms. A selection variable which is labeled as *MR* was included to correct for the sample selection bias. The negative and significant parameter estimates showed the presence of selectivity in investment decisions. The estimates show whether and to what extent the firms invest in innovative input, once they are engaged in innovative activities.

The coefficient of sales and size are significant, but show opposite directions. It is somewhat surprising that our results show that the R&D intensity decreases significantly with the firm size in terms of the number of employees. However the dependent variable in this model is R&D investment per employee, that the increasing number of employees will decrease the investment per employee by simple calculation. The number of patents pending also increases R&D investment per employee of firms. The firms included in the petroleum-based products, chemical compounds, plastic and rubber industries, machinery industry, computer, business machine, electrical machinery, electrical components and precision industries, automobile and transportation manufacturing industries, showed a higher level of R&D investment per employee, compared to the food and tobacco industries. The industries listed above are namely heavy industries and information technology industries, which require much capital input and R&D per employee. The firms in Jeon-la province and Gwang-ju city showed significantly low R&D intensity compared to firms located in Gyung-gi province, Seoul and In-chon.

Table 2. Heckman's Two Step (Probit/OLS) Estimation method

PROBIT				OLS			
Variables	Parameter	STD	t-test	Variables	Parameter	STD	t-test
Intercept	-2.29	1.13	**	Intercept	2.11	2.17	
SIZE	0.44	0.10	***	SIZE	-0.30	0.16	*
SALES	0.12	0.07	*	SALES	0.39	0.10	***
EXPORT	0.02	0.01		EXPORT	-0.01	0.01	
WAGE	-0.14	0.07	**	CAPITAL	0.11	0.08	
FEQUITY	0.04	0.02	**	PATENT	0.29	0.07	***
MERGER	0.20	0.20		MERGER	0.30	0.26	
CLOSING	-0.29	0.25		CLOSING	-0.14	0.36	
VENTURE	0.95	0.12	***	VENTURE	0.41	0.31	
TRANSFER	0.23	0.11	**	TRANSFER	0.07	0.15	
CFS	0.00	0.05		MR	-1.11	0.63	*
CISK	-0.01	0.05		CFS	0.00	0.07	
CESK	0.19	0.05	***	CISK	-0.20	0.07	***
CLCO	-0.03	0.05		CESK	0.08	0.09	
CICO	0.11	0.05	**	CLCO	-0.16	0.06	**
CHPF	0.03	0.05		CICO	0.06	0.07	
CHIF	0.09	0.05	*	CHPF	-0.01	0.07	
SIZE2	-0.05	0.14		CHIF	-0.02	0.06	
SIZE3	-0.23	0.23		SIZE2	-0.10	0.19	
IND2	0.04	0.27		SIZE3	-0.21	0.29	
IND3	-0.71	0.32	**	IND2	0.35	0.41	
IND4	0.25	0.23		IND3	0.72	0.66	
IND5	-0.08	0.23		IND4	0.71	0.35	**
IND6	0.09	0.26		IND5	0.47	0.35	
IND7	0.29	0.24		IND6	1.14	0.37	***
IND8	0.22	0.24		IND7	1.53	0.35	***
IND9	-0.04	0.36		IND8	1.15	0.36	***
AREA2	-0.04	0.14		IND9	0.22	0.54	
AREA3	0.30	0.20		AREA2	-0.15	0.20	
AREA4	-0.12	0.11		AREA3	-0.74	0.27	***
AREA5	-0.49	0.38		AREA4	-0.22	0.16	
				AREA5	-0.53	0.71	
Number of Observation	1030			Number of Observation	638		
Log Likelihood	-545.01			Adj. R-square	0.30		

***, **, and * indicates 1%, 5%, and 10% significance level, respectively.

6.2. The Innovation Output Results

The system of innovation output and productivity equation was estimated and the results are presented in Table 3. The two equations are estimated as a system of interdependent equations by the 3SLS method.

Table 3. 3SLS system of equations

Dependent Variable	Innovation Output			Dependent Variable	Growth in Labor Productivity		
Variables	Parameter	STD	t-test	Variables	Parameter	STD	t-test
Intercept	-0.04	0.34		Intercept	1.97	0.39	***
INNOVIN (predicted)	0.00	0.10		INNOVOUT (predicted)	0.21	0.05	***
PROD (growth, predicted)	0.26	0.09	***	PROD (lagged)	-0.22	0.02	***
SALES	0.01	0.05		SIZE	0.07	0.03	**
FEQUITY	0.00	0.00		CAPITAL	0.01	0.02	
MR	0.15	0.10		EXPORT	0.00	0.00	
MERGER	0.08	0.07		MERGER	-0.02	0.06	
CLOSING	0.05	0.08		CLOSING	-0.28	0.09	***
VENTURE	0.11	0.10		VENTURE	-0.05	0.04	
TRANSFER	0.09	0.03	***	TRANSFER	-0.02	0.03	
PATENT	0.05	0.03	*	PATENT	0.02	0.02	
PROCINNOV	0.09	0.03	***	CFS	0.00	0.02	
CFS	0.04	0.02	**	CISK	-0.01	0.02	
CISK	0.05	0.02	*	CESK	-0.01	0.02	
CESK	0.05	0.02	**	CLCO	-0.02	0.02	
CLCO	0.01	0.02		CICO	0.01	0.02	
CICO	0.02	0.02		CHPF	0.00	0.02	
CHPF	0.01	0.01		CHIF	-0.01	0.01	
CHIF	0.03	0.01	**	SIZE2	-0.09	0.05	**
SIZE2	-0.03	0.04		SIZE3	-0.13	0.07	*
SIZE3	-0.03	0.07		IND2	-0.12	0.10	
IND2	0.08	0.09		IND3	0.05	0.15	
IND3	-0.02	0.13		IND4	0.11	0.08	
IND4	-0.04	0.11		IND5	0.05	0.09	
IND5	-0.09	0.09		IND6	-0.07	0.09	
IND6	0.04	0.14		IND7	-0.15	0.09	*
IND7	0.13	0.18		IND8	-0.04	0.09	
IND8	-0.04	0.14		IND9	0.03	0.13	
IND9	0.03	0.11		AREA2	-0.03	0.05	
AREA2	0.01	0.05		AREA3	0.11	0.07	*
AREA3	0.01	0.09		AREA4	-0.01	0.04	
AREA4	0.02	0.04		AREA5	0.17	0.17	
AREA5	-0.12	0.17					
Number of Observation	638						
System Weighted R-square	0.267						

***, **, and * indicates 1%, 5%, and 10% significance level, respectively.

The results show that the coefficient of innovation output in the labor productivity equation and the coefficients of labor productivity growth in the innovation output equations are highly statistically significant. This suggests a two-way causal relationship between innovation output and labor productivity growth and in support of 3SLS as an appropriate estimation method. The relationship between innovation output and predicted level of innovation input was insignificant.

Innovation output, defined as the ratio of sales originated from innovative products to the total sales, have a positive and significant relationship with predicted labor productivity growth, technology transfer, the number of patents pending and process innovation. However, we could not find evidence of heterogeneity in the innovation output by size of firms, industrial sector and location of firms.

6.3. The Labor Productivity Growth Results

The results from the labor productivity growth presented in Table 3 show that growth in labor productivity increases largely with knowledge capital. That is, the elasticity of innovation, which is predicted innovation output, obtained by controlling other variables on innovation output to labor productivity was positive. The lagged level of labor productivity shows negative and significant parameter estimates which suggest a converging level of labor productivity over time. A similar result has also been found from descriptive analyses of Oh et al. (2006).

The variables indicating firm size show somewhat puzzling results. The size variable is positively correlated with growth of labor productivity, however, size dummies shows that as firms are located in the larger size classes, their productivity growth is hampered. So it is true, that larger firms grow faster in terms of labor productivity, but the tendency decreases as firms get bigger. Sales decrease more than 10 percent due to closing some of the plants found to be negatively significant to growth of labor productivity. Looking at the estimates for the industrial dummies, we find that firms in computer, business machine, electrical machinery, electrical components and precision industries showed a significantly lower level of labor productivity growth, compared to firms in the food and tobacco industry.

7. Summary and Conclusions

This study aimed to investigate the relationship between innovativeness and productivity of Korean manufacturing firms. In doing so, we have examined the possible two-way causal relationships between knowledge capital and productivity among the Korean innovative manufacturing firms. We have used a knowledge production function approach formulation of the link between incentive to invest in innovation activities, innovation input, innovation output and labor productivity growth by Pakes and Griliches (1984). The models refined by Crépon et al. (1998) are estimated by adopting a multi-step econometric approach suggested in Lööf and Heshmati (2002 and 2006).

For the empirical analysis we have used Korean Innovation Survey data released in 2002, which contains detailed information on various key characteristics of Korean manufacturing firms related to innovation characteristics and KISinfo, which contains the panel data of financial statements of firms. We controlled for firm specific variables that are related to firms' innovation activities such as firm characteristics and innovation environmental variables. The main findings are summarized as: (i) larger firms tend to invest more in R&D, (ii) firms with higher foreign equity are engaged in R&D activity more often, (iii) firms having more patents pending invested more in R&D, (iv) productivity growth increases innovation output, (v) it is hard to find evidence that increasing investment in R&D directly increases innovation output, (vi) the relationship between the ratio of innovation output to the total sales and size of firm is insignificant, (vii) firms doing technology transfer and having more patents have more innovation output, (viii) increased amount of innovation output also facilitates productivity growth and (ix) larger firms grow faster in terms of productivity growth, however, as their size gets bigger, the tendency is slowed.

The listed results suggest that there is a two-way causal relationship between innovation output and productivity growth. The current policy to enhance the productivity of the manufacturing sector needs more complicated instruments other than just letting firms to invest more in R&D. An actual linkage from R&D investment to innovation output was not established. Policy makers must reflect on this issue in depth before designing policy measures to enhance productivity.

However, this research has some limitations concerning the data, because panel data of the innovation survey has not been made available yet. It is difficult to establish causal relationships in a cross sectional framework and under the strong assumption of constant stream of innovation investment and innovation output over time. Innovation is a process which takes a long period of time and just glancing at the cross-sectional process does not give many opportunities to go deeper into the process. If a Korean innovation survey is continuously conducted and accumulated in a proper fashion, then it will be able to give more information on this topic.

APPENDIX 1. THE DEFINITION OF THE DEPENDENT AND INDEPENDENT VARIABLES

Variables	Definition of the variables
INNOV	Dummy variable indicating whether the firm has positive R&D expenditure in 2001
INNOVIN	Log of R&D expenditure per employee in 2001
INNOVOUT	The ratio of innovative sales to total sales
PROD	Log of value added per employee. The value added was calculated by adding gross profit and labor cost, as in Miller (1986).
SALES	Log of sales per employee in 2001
SIZE	Log of number of employees in 2001
EXPORT	Log of amount of export per employee in 2001
PROFIT	Log of profit per employee in 2001
WAGE	Log of average wage level in 2001
CAPITAL	Log of capital per employee in 2001
FEQUITY	The percentage of foreign equity ownership, averaged value for year 2001
PATENT	Log of number of patent pending during 2000 and 2001

Appendix 1. (Continued)

Variables	Definition of the variables
MERGER	Dummy variable indicating whether a merging in 2000-1 increased total sales by 10% comparing 1999 and 2001
CLOSING	Dummy variable indicating whether exit during 2000-2001 increased total sales by 10% comparing 1999 and 2001
VENTURE	Dummy variable indicating whether the firm is certified by the government as a venture business
TRANSFER	Dummy variable indicating whether a firm has experienced technology transfer from external sources.
PROCINNOV	Process innovation
CFS	Composite index about firm strategy
CISK	Composite index about Internal source of knowledge
CESK	Composite index about external source of knowledge
CLCO	Composite index about location of cooperation
CICO	Composite index about importance of cooperation
CHPF	Composite index about hampering project factors
CHIF	Composite index about hampering innovation factors
SIZE1	Number of employees; less then 85
SIZE2	Number of employees; 85 – 179
SIZE3	Number of employees; 180 or more
AREA1	Gyung-gi province, Seoul and In-chon
AREA2	Chung-cheong province and Daejon
AREA3	Jeon-la province and Gwang-ju
AREA4	Gyeong-sang province, Busan, Daegu and Ulsan
AREA5	Jeju province and Gang-won province
IND1	Food and tobacco industries
IND2	Fiber, textile, clothes, bags and shoes industries
IND3	Wooden products, pulp, paper, publication and press industries
IND4	Petroleum based products, chemical compounds, plastic and rubber industries
IND5	Base metal industry, ferrous metal, non-ferrous metal and metalworking industries
IND6	Machinery industry
IND7	Computer, business machine, electrical machinery, electrical components and precision industries
IND8	Automobile and transportation manufacturing industries
IND9	Furniture and reproductive industries

Notes: Before taking log of variable, a number one is added to each variable to prevent taking log of zero.

REFERENCES

Aghion, P. and Howitt, P. (1998), *Endogenous Growth Theory*, MIT Press, Cambridge, MA.

Arrow, K. J. (1962), The economic implications of learning by doing, *The Review of Economic Studies*, 29, 155-173.

Baldwin, J. R. and Hanel, P. (2003), *Innovation and Knowledge Creation in an Open Economy*, Cambridge University Press, 2003.

Cohen, W. M. and Klepper, S. (1996), A reprise of size and R&D, *The Economic Journal*, 106, 925-951.

Cohen, W. M. and Levinthal, D. A. (1989), Innovation and learning: the two faces of R&D, *The Economic Journal*, 99, 569-596.

Crépon, B., Duget, E. and Mairesse, J. (1998). Research, innovation, and productivity: An econometric analysis at the firm level, NBER Working Papers 6696.

Evangelista, R., Perani, G., Rapti, F. and Archibugi, D. (1997), Nature and impact of innovation in manufacturing industry: some evidence from the Italian innovation survey, *Research Policy*, 26, 521-536.

Griliches, Z. (1988), Hedonic price indexes and the measurement of capital and productivity: some historical reflections, NBER Working Papers 2634.

Griliches, Z. (1990), Patent statistics as economic indicators: A survey, *Journal of Economic Literature*, 28, 1661-1707.

Griliches, Z. (1994), Productivity, R&D and the data constraint, *American Economic Review*, 84 (1), 1-23.

Hall, B. H. and Mairesse, J. (1995), Exploring the relationship between R&D and productivity in French manufacturing firms, *Journal of Econometrics*, 65, 263-293.

Heshmati, A. (2006), A generalized knowledge production function, RATIO Working Papers No. 89, The RATIO Institute, Sweden.

Korea Productivity Center (2004), International comparisons of productivity level, Korea Productivity Center.

Klette, T. J. and Griliches, Z. (1997), Empirical patterns of firm growth and R&D investment: A quality ladder model interpretation, NBER Working Papers 5945.

Lööf, H. and Heshmati, A. (2002), Knowledge capital and performance heterogeneity: A firm-level innovation study, *International Journal of Production Economics*, 76, 61-85.

Lööf, H. and Heshmati, A. (2006), On the relationship between innovation and performance: A sensitivity analysis, *Economics of Innovation and New Technology*, 15, 317-344.

Mairesse, J. and Sassenou, M. (1991), R&D productivity: A survey of econometric studies at the firm level, NBER Working Papers 3666.

Miller, E. M. (1986), Determinants of the size of the small business sector: They are labor productivity, wage rates and capital intensity, *American Journal of Economics and Sociology*, 45, 389-402.

Oh, I., Heshmati, A., Baek, C. and Lee, J-D. (2006), Comparative Analysis of Firm Dynamics by Size: Korean Manufacturing, RATIO Working Papers No. 94, The RATIO Institute, Sweden.

Pakes A. and Griliches, Z. (1984), Patents and the R&D at the firm level: a first look, in: Z. Griliches (ed.), *R&D, patents and productivity*, Chicago University Press, Chicago.

Ranga Chand, U. K. (1982), Characteristics of research and development performing firms in Canadian manufacturing, *Research Policy*, 11, 193-203.

Shin, T. Y., Song, W. J., Um, M. J. and Lee, J. Y. (2002), Korean innovation survey 2002: Manufacturing sector, Science and Technology Policy Institute (www.stepi.re.kr), Korea.

Solow, R.M. (1997), *Learning from "Learning by Doing": Lessons for Economic Growth*, Stanford University Press, Stanford, CA.

Zhang, A., Zhang, Y. and Zhao, R. (2003), A study of the R&D efficiency and productivity of Chinese firms, *Journal of Comparative Economics*, 31, 444-464.

In: Micro-Evidence for the Dynamics of Industrial Evolution ISBN: 978-1-60456-388-7
Editors: A. Heshmati, J.-D. Lee

Chapter 8

THE THRESHOLD EFFECT ON THE GROWTH OF SMEs IN KOREAN MANUFACTURING

Chulwoo Baek [a,*] ***and Jeong-Dong Lee*** [b,†]
[a]Korea Institute of Science and Technology Evaluation and Planning, Seoul, Korea
[b]Seoul National University, Seoul, Korea

ABSTRACT

There is strong concern about intentional retardation in the growth of SMEs (Small and Medium sized Enterprises). This is supported by the fact that more than 5 billion dollars are given to SMEs exclusively, but 50 regulations are imposed on LSEs (Large sized enterprises).

In contrast to the previous claim based on questionnaires and case studies, as part of our research described in this paper, we conducted a regression analysis with the database of all statutory audit firms in order to verify the existence of a threshold effect. Empirical results showed that the existence of a threshold effect cannot be verified in this analysis framework.

Keywords: Threshold effect; SME; Statutory audit firm; Manufacturing
JEL: D21, D92, L52

1. INTRODUCTION

SMEs (Small and Medium sized Enterprises) are given much weight in Korean manufacturing in that they constitute 99.8% of the entire population of enterprises in Korean manufacturing (3 million SMEs) and 86.5% of the total employment (10.41 million

* email contact: torke@paran.com
† email contact: leejd@snu.ac.kr

employees) as of 2004.1 With respect to employment, during the period from 1999 to 2004, the number of those employed by LSEs (Large sized Enterprises) was reduced by 1.2 million while that of SMEs increased by 1.54 million (Korean Small and Medium Business Administration). This status of SMEs is also common in advanced economies. SMEs, in most advanced economies, constitute more than 95% of the establishments and 2/3 of employment (OECD, 2005).

On the basis of its importance, 'support for SMEs' has been one of the mainstream industrial policies in Korea for several decades. In 2005, the policy fund for SMEs amounted to about 5 billion dollars which was about 5% of the government budget (National Assembly Budget Office, 2006). Besides this direct grant, SMEs are given 13 exclusive tax incentives in the various types of tax deductions (The Federation of Korean Industries, 2006).

Firms are analogous to living creatures, meaning that they should maintain an appropriate growth rate for their survival in fierce competition. Even though an optimal growth rate can be differentiated according to production capability and market size, it cannot be denied that continuous growth is an unavoidable survival strategy.

However, there is concern about the disconnection between SMEs and LSEs. This is called 'the dilemma of 300 employees'. Once a SME grows into a LSE, all policy support for the former SMEs is deprived simultaneously with a period of grace of three years. Therefore, even though a firm has enough capability to grow into an LSE, the firm hesitates to cross the bridge into a LSE (Economist, 2005). According to Kim's research (2005b), only 75 establishments among the 56,472 SMEs in 1993 have grown into LSEs in 2003.

Debate about this phenomenon has been raised mainly by The Korea Chamber of Commerce and Industry (KCCI) and The Federation of Korean Industries (FKI). The KCCI (2006) pointed out that SMEs intentionally retard their growth when their number of employees is in the vicinity of close to but less than 300. It is suggested that SMEs feel the burden from the fact that marginal profit from additional employment may not be more than the marginal loss caused by deprivation of government support. Instead, as an expedient, SMEs sometimes choose to split themselves or establish an affiliated firm.

In the same vein, FKI (2004) insisted that regulations on LSEs, which were introduced in order to suppress the concentration of economic power, expanded out of the area of the fair trade act. Such discriminatory regulations according to firm size have checked the natural growth of firms by contracting investment.

The above research findings consistently assert the following factors. Firstly, it is certain that the threshold effect exists in the vicinity of less than 300 employees. Secondly, therefore, in order to promote smooth growth from an SME to an LSE, policy support should be maintained to some extent even when the number of employees exceeds 300 or regulations for LSEs need to be mitigated.

However, an academically strict analysis has never been conducted to identify a threshold effect for SME growth. Previous claims were mainly based on case studies or questionnaires for specific firms; hence, these results have limitations of generalization for the whole economy. Therefore, this research attempted to verify the existence of a threshold effect on SME growth by applying regression analysis to all of the statutory audit firms. This result is

---

[1] SME is legally defined and each industry sector has a different standard. In the case of the manufacturing sector, the firms whose regular employees are less than 300 or whose equity capital is 8 million dollars or less are designated as SMEs.

expected to offer a basis of debate about whether present industry policy, mainly oriented to SME support, should be changed or not.

The rest of this paper is organized as follows. Policy support for SMEs and regulations for LSEs are explained in Section 2. Section 3 reviews previous literature on the threshold effect and Section 4 explains the regression model. Section 5 describes the data set and variables used in the analysis. Section 6 estimates the threshold effect. Finally, Section 7 presents concluding remarks.

2. Institutional Setting

2.1. Policy Support on SMEs

'Support for SMEs' has been the main motto of the government in industrial policy. The Small and Medium Business Administration (SMBA) was established in 1996 for the purpose of systematic and efficient support for SMEs. Fourteen government departments are involved in support of the policy. Policy support for SMEs mainly consists of a policy fund and tax incentives.

First, the policy fund, such as R&D support and financing support, is executed as a type of subscription, subsidy, loan, and investment. In 2005, 204 programs from 14 government departments were performed. The total amount of policy funds for SMEs was about 5 billion dollars (National Assembly Budget Office, 2006).2

As shown in Table 1, more than 70% of the policy fund is executed as a type of loan which requires repayment. Subscriptions and subsidies are usually offered to a public organization dealing with a credit guarantee. An investment is made by acquiring a share of an investment fund for SMEs' support.

Table 1. Policy fund for SMEs
(unit: million dollars, %)

Year	2002	2003	2004	2005
Loan	3,269	3,867	3,624	4,205
	(74.5)	(76.9)	(73.2)	(82.7)
Subscription	900	1,0141	1,190	704
& Subsidy	(20.5)	(20.2)	(24.0)	(13.8)
Investment	216	145	138	179
	(4.9)	(2.9)	(2.8)	(3.5)
Total	4,385	5,026	4,953	5,088
	(100)	(100)	(100)	(100)

Source: National Assembly Budget Office (2006).

[2] The total amount of the policy fund given to SMEs was about 7.3 billion dollars in 2005. However, this amount included the policy fund which was consequently given to the SMEs, although it allowed for the LSE to be a beneficiary. 5 billion won (Korean currency) is the total money amount of programs whose proportion to SMEs is more than 90%.

These funds are used for the installation of equipment, operating expenses, R&D, and so on. Given that most SMEs suffer from a shortage of financial resources, a policy fund which offers favorable treatment in interest, mortgage, and an exchange period is very attractive to SMEs.

Secondly, a tax incentive as well as a support fund offers significant benefit for SMEs. Table 2 categorizes the tax incentives for SMEs into tax credit for investment, a reserve for investment, tax incentives for newly established SMEs, and special tax incentives for SMEs.

2.2. Regulations for the LSEs

Regulations for the LSEs were introduced in 1987 in order to suppress the concentration of economic power. Its application scope was expanded in the 1990s. According to asset size, differentiated regulations are applied as shown in Table 3. A total of 50 regulations are enforced for LSEs. These regulations are based on various legal acts, such as the fair trade act, the external audit act, and the securities exchange act. The most remarkable characteristic in the regulations is that more than 3/4 of the regulations are enforced where the firm's asset size exceeds 2 billion dollars.

Table 2. Tax incentives on SMEs

Tax credit for investment
If SMEs acquire business assets such as machinery and equipment or installation of an information management system, 3% of the acquisition amount is deducted from income tax or corporation tax.
Reserve for investment
The reserves are treated as losses in calculating income within the limit of 20% of the value of the business assets. The aforementioned reserves used as expenditures for investment shall be added back as gains in each taxable year in 36-month installments from the business year to which the third year belongs.
Tax incentives for newly established SMEs
When new SMEs are established in areas other than Seoul or its adjacent areas, the corporation tax for such businesses is reduced by 50% for the first six years. The property tax on business assets belonging to new SMEs is reduced by 50% for five years after establishment. The acquisition tax and the registration tax are reduced by 100% for two years.
Special tax incentive for SMEs
SMEs are eligible for a 30% reduction in corporation tax if located in non-metropolitan areas. Small sized enterprises located in metropolitan areas are eligible for 20% reduction.

Source: Son, 2002, Corporate Income Tax in Korea.

Table 3. Regulation on LSEs according to asset size

Regulation threshold	Main regulation	No. of relevant acts	No. of regulations
Asset size of more than 5 billion $	The ceiling on equity investment Prohibition against participation in animal husbandry Exclusion of support from employment protection legislation	5	5
Asset size of more than 3 billion $	Prohibition against the possession of stocks of broadcasting companies	1	2
Asset size of more than 2 billion $	Prohibition against cross investment Restriction on establishment of a holding company Limitation on voting rights of financial affiliates Exclusion from application of succession donation Obligation formation of external board of directors Obligation formation of audit committee	17	31
Asset size of more than 100 million $	Obligatory designation of external audit Obligatory designation of full-time audit	4	5
Credit grant	Integrated supervision of main creditor bank on high-debt firms	1	1
Designated large conglomerate	Rejection of unfair act and calculation	3	3
LSEs or equity size of more than 100 million $	Exclusion of interest expenses to loss in money	3	3
Total		**25[a]**	**50**

Source: The Federation of Korean Industries (2004).

[a] Overlapping calculation of relevant acts.

In addition, the LSEs have a disadvantage in taxation. When their equity exceeds 100 million dollars, interest expenses of loans in excess of the standard are not included as loss of money. The investment in a nonprofit foundation can be deducted from inheritance tax and donation tax; however, the firms, whose asset value is more than 2 billion dollars, are excluded from this tax benefit.

3. Previous Literature on the Threshold Effect

The threshold effect has usually been dealt with in the relationship between employment protection legislation (EPL) and the growth of firms. EPL requires an additional firing cost when the number of employees arrives at a certain level; therefore, there is the possibility for firms to retard their growth intentionally so as not to exceed the regulation threshold.

Schivardi and Torroni (2004) analyzed an Italian case where the EPL is substantially less stringent for a firm with less than 16 employees. In order to measure the effect of the threshold on growth propensity, they run a probit of positive employment change, in size and additional controls, on a polynomial. The result showed that the regression model fits the data remarkably well except in the proximity of the threshold. The probability the growth of firms in the proximity of the threshold is reduced by around 2 percent points.

For the same Italian case, Borgarello et al (2002) characterized the threshold effect as an increase of inaction, and a reluctance to grow. They estimated the persistence probability and the difference between the probability of increasing and decreasing firm size. As a result, they verified that firms close to the 15 employees threshold experience an increase in persistence of 1.5% with respect to a baseline statistical model. Firms with 15 employees are more likely to move backwards than upwards.

Verick (2004) conducted an empirical study of a German case involving the Protection Against Dismissal Act (PADA). The PADA provides protection to workers from unfair dismissal where the unit regularly employs more workers than the threshold for applicability of the legislation. However, the threshold for applicability of the PADA has changed twice, once from 5 to 10 workers in 1996 and then back again to 5 workers in 1999. The effects of the changes to the PADA are estimated using the difference-in-differences (DID) estimator. The results show that a tightening of the threshold results in a higher probability of persistence and a lower probability of growth.

Acemoglu and Angrist (2001) dealt with the case involving the Americans with Disability Act (ADA). The ADA requires employees to accommodate disabled workers and outlaws discrimination against the disabled in hiring, firing, and pay. As a result, although the number of disabled individuals receiving disability transfers increased at the same time, the decline in employment of the disabled does not appear to be explained by increasing transfers alone leaving the ADA as a likely cause.

Out of the issues in employment protection, Denkabe (2003) examined the contribution of foreign aid to economic growth in the context of macroeconomic policy by the use of a dynamic growth equation which was estimated by way of the Generalized Method of Moments. He suggested the existence of a threshold value of aid, defined by macroeconomic policy, below which aid tended to have a positive effect on economic growth and beyond which diminishing returns to aid may generate a non-positive impact on growth.

From previous literature, the following facts can be drawn. First, the threshold effect has been analyzed mainly in relation to employment protection and is a significant research theme in policy study. However, up to now, there has not been much study of the threshold effect related to other issues except for the research of Denkabe (2003).

Second, although each previous research effort adopted a different regression model, the capture of the threshold effect can be detected by the abnormal growth pattern in a region of less than the threshold value.

4. MODEL

The purpose of this research is to verify whether or not firms intentionally retard their growth in the vicinity of less than 300 employees. Therefore, the start point of the discussion is how to identify the threshold effect in data. In this sense, we choose two yardsticks – the possibility of decreasing the size and the growth rate.

The former variable is expected to indicate whether firms are likely to decrease their size in order to maintain current policy support; thus, the pattern of size decrease shows an abnormal pattern from that in other size ranges (Verick, 2004; Schivardi and Torrini, 2004). This variable is defined as 1 when a firm decreases the number of employees compared with last year; otherwise, this variable is defined as 0.

On the other hand, to decrease the number of employees can be an extreme strategy. In reality, firms may choose a strategy to reduce the growth rate intentionally in order to slow down the time of becoming an LSE rather than to decrease the number of employees. In this sense, the second variable, the growth rate, can be a good alternative which can be complementary to the possibility of decreasing.

If these two variables show an abnormal pattern in the region of less than 300 employees, we can verify the existence of a threshold effect. For econometric analysis, the regression model can be constructed as Equation (1).

$$Y_{it} = \beta_o + \beta_X X_{it-1} + \beta_W W_t + \beta_Z Z_i + \beta_D D(size) \qquad (1)$$

where, β_0 : intercept

X_{it-1} : characteristics of firm i at year $t-1$

W_t : unemployment rate at year t

Z_i : dummy variable for the industry sector

$D(size)$: dummy variable for the firm size

As aforementioned, Y_{it} can be defined as in Table 4. When Y_{it} is the possibility of decreasing, the dependent variable becomes binary. Therefore, a probit regression model is the most suitable (Schivardi and Torroni, 2004). Meanwhile, in the case of the growth rate, the dependent variable is a continuous value, therefore we adopted an OLS regression model. We expected that these two regression models could identify the threshold effect from a different but complementary perspective.

X_{it-1} is a control variable for considering the differences in characteristics among firms. It consists of the natural log of employees, firm age, and operating income over the revenue ratio at year $t-1$. Given the apparent non-linearity between the dependent variable and firm size, we have taken a fourth-degree polynomial in size. Age was introduced in the standard quadratic form (Schivardi, 2004).

Table 4. The definition of Y_{it} and relevant regression models

Y_{it}	Definition	Regression model
The possibility to decrease the size	If the number of employees at year t is larger than at year t-1, then 1. Otherwise, 0	Probit
The growth rate	The increase ratio of employees at year t to employees at time t-1 (%)	OLS

W_t is a control variable for the macroeconomic effect. Generally the GDP growth rate is used. However, the dependent variable is related to employment, we adopted the unemployment rate for controlling the macroeconomic effect.3 In addition, whether a certain industry belongs to a mature industry or growing industry can have an effect on the dependent variable; therefore, we inserted the dummy variables of the industry sector (Z_i) at the Standard Industry Classification (SIC) code 2-digit in order to control the industry effect. $D(size)$ is introduced in order to detect the abnormal change of the dependent variable. $D(size)$ is given at a 50 employees interval from 100 to 399 (D(200~249), D(250~299), D(300~349)) and is given at only D(400~) for more than 400 employees. The existence of a threshold effect can be verified by the coefficient of D(250~299) showing a different pattern from D(200~249) and D(300~349).

5. DATA

The credibility of the empirical study is linked directly to the scope and accuracy of the database used. More specifically, the database should cover the complete set of firms in the relevant industry sector so as to be as comprehensive as possible in order to draw a policy implication for the economy from the empirical results. Moreover, the numerical value in the database should be reliable in accuracy. Its verification by an external organization would improve the accuracy.

The Mining and Manufacturing Census (MMC) surveyed by the Korea National Statistical Office is the most appropriate database in that it covers the total number of establishments with 5 or more employees. However, the MMC contains not firm-level data but establishment-level data. This means that the size of multi-plant firms can be underestimated.4 If we consider that most of the multi-plant firms are LSEs, then the MMC database is an inappropriate database in analyzing the growth pattern of firms in the vicinity of less than 300 employees.

---

[3] We also inserted the GDP growth rate as a control variable for the macroeconomic effect instead of the unemployment rate. In this case, the regression result is almost the same as the one with the unemployment rate.

[4] Ahn (2005) analyzed the growth of firms with MMC. However, he also mentioned this database is likely to underestimate the size of multi-plant firms.

In this sense, we used unbalanced panel data from the Korea Information System (KIS). The KIS offers data on the financial statements of all statutory audit firms. The firms whose asset size is more than the legal standard should have an audit at the end of each financial year. The legal standard has changed as shown in Table 5. Most of the firms in the vicinity of less than 300 employees are statutory audit firms, therefore the use of data of whole statutory audit firms corresponds to the purpose of this research. Auditing by external accounting firms improves the credibility of data. Moreover, the data of the financial statement has been widely used in empirical studies (Chen et al, 2003; Lee, 2003).

This research confined its target to manufacturing firms. Because the service industry, such as finance and distribution, has a different growth pattern and the elasticity of labor from manufacturing firms, simultaneous analysis in the same pool can result in a distorted result.

Table 6 represents a data description of the KIS data. The number of observations has trebled since 1990; however, the mean and standard deviation of employees shows a rapid decrease. This implies that in spite of the increasing legal standard, the number of the statutory audit firms increased and this additional statutory firm is smaller than the average of the incumbent statutory audit firms at a previous year.

Table 5. Legal standard for a statutory audit firm

Time	Legal standard of asset size (million dollars)
1980 (Beginning)	3 (or equity size of more than 500 thousand dollars)
April 1988	3
March 1990	4
May 1993	6
April 1998	7

Table 6. Descriptive statistics of the dataset

Year	No of obs.	No. of employees	
		Mean	Standard deviation
1990	2,155	582	1,856
1991	2,451	517	1,765
1992	2,547	484	1,705
1993	2,571	474	1,735
1994	2,942	443	1,797
1995	3,422	401	1,745
1996	3,564	386	1,744
1997	4,021	333	1,532
1998	4,120	284	1,295
1999	4,698	265	1,311
2000	5,096	249	1,284
2001	5,816	232	1,194
2002	6,434	218	1,175
2003	6,700	218	1,227
2004	6,304	232	1,326
Total	62,841		

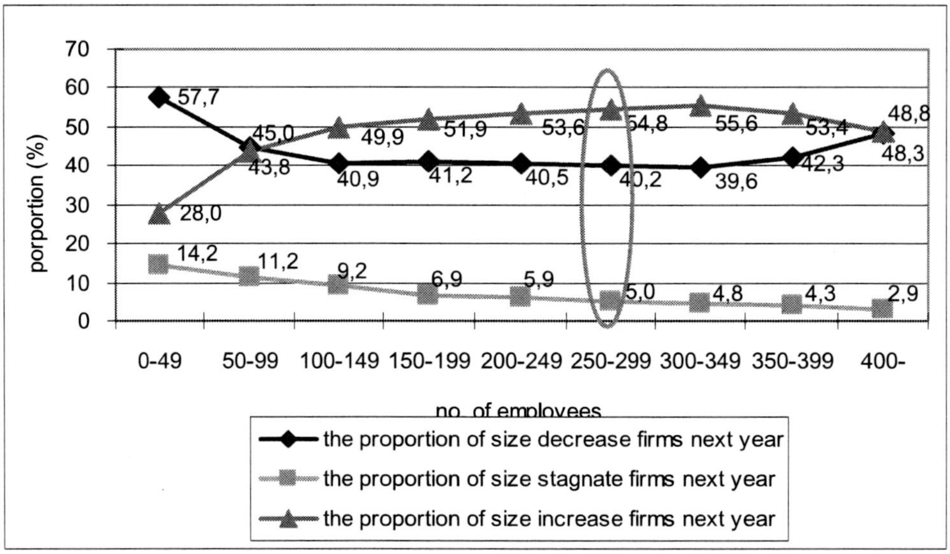

Figure 1. The proportion of firms whose employees increase/stagnate/decrease next year.

6. Empirical Results

Prior to the econometric analysis, we conducted a preliminary test. Figure 1 shows the proportion of firms whose number of employees increases, stagnates, and decreases in the following year without consideration of heterogeneity among firms and the macroeconomic effect. In the case of the proportion of firms with size increases, the firms in the range of 50~99 employees had a 15% point larger value than the firms in the range of 0~49 employees. This proportion shows continuous increasing patterns in the range of 300~349 employees, then it declines. The proportion of firms decreasing in size the following year indicates the exact opposite tendency against that of firms increasing in size. The proportion of firms reducing in size the following year is slowly declining. Generally, the larger the size is, the higher the possibility to increase in size. Aforementioned results, without consideration of other effects, imply that the growth pattern of firms in the range of 250~299 employees does not deviate from that of firms in the range of 200~249 and 300~349, which means the non-existence of a threshold effect in the vicinity of less than 300 employees.

Next, we conducted a regression analysis with consideration of other effects, such as heterogeneity among firms, industrial differences, and macroeconomic effect. As explained in Table 4, model 1 adopted a probit regression whose dependent variable was the possibility of decreasing in size. Model 2 used OLS regression whose dependent variable was the growth rate. Dummy variables of the industry sector were inserted at the Standard Industry Classification (SIC) 2-digit code in order to control the industry effect.

According to model 1 in Table 7, the coefficients of the size dummies have a tendency to increase. This means that as the firm size grows, the propensity to decrease the number of employees is larger. Model 2 shows that the larger the firm size is, the smaller the growth rate. Based on the regression, the possibility of decreasing the size and the growth rate relative to those of the firms with less than 100 employees is represented in Figure 2.

Table 7. Regression results

		Model 1 (Probit)		Model 2 (OLS)	
	intercept	-0.7459	(***)	10.616	(***)
Firm Characteristics	log(lag worker)	0.8039	(***)	-7.2938	(***)
	log(lag worker)^2	-0.1408	(***)	1.7820	(***)
	log(lag worker)^3	-0.0041		-0.1822	(***)
	log(lag worker)^4	0.0011	(***)	0.0067	(***)
	age	-0.0041	(***)	-0.0040	(***)
	age^2	-0.0000		0.0001	(***)
	operating income over revenue ratio	0.0001		0.0001	
Size Dummy	dummy(100-149)	0.2633	(***)	-0.0739	(***)
	dummy(150-199)	0.3769	(***)	-0.1505	(***)
	dummy(200-249)	0.4765	(***)	-0.2173	(***)
	dummy(250-299)	0.5238	(***)	-0.2668	(***)
	dummy(300-349)	0.5809	(***)	-0.3107	(***)
	dummy(350-399)	0.6267	(***)	-0.3321	(***)
	dummy(400-)	0.8141	(***)	-0.4137	(***)
Macroeconomic effect	unemployment rate	-0.0237	(***)	-0.0208	(***)

*, ** and *** indicate statistical significance at the 10, 5 and 1% levels, respectively. The dummy variables of the industry sectors are inserted at KSIC code 2 digit, however, their coefficients are not statistically significant and thus not represented in the table.

We can conclude from Table 7 and Figure 2 that as firms grow, they have a lower possibility of decreasing their size with higher growth rate. More importantly, the abnormal pattern in the vicinity of less than 300 employees is not observed from the perspective of the possibility of decreasing the size and the growth rate. This means that the aforementioned, the dilemma of 300 employees, is not verified in our analysis framework..

The reasons why the threshold effect, which is asserted to be true in many previous reports and articles, is not identified in this research can be explained as follows. First, we can imagine the case that even though the threshold effect actually exists, the data coverage prevents us from capturing this effect. Although the data of all the statutory audit firms is the best data we can choose, it is likely to cover most LSEs with more than 300 employees, but omit some SMEs. Therefore this may cause the sample bias in this analysis.

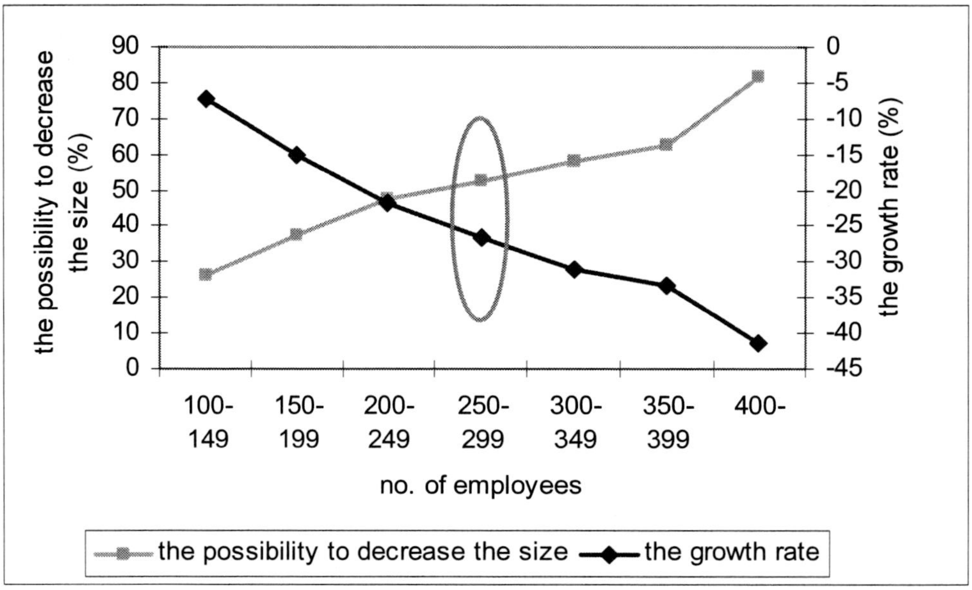

Figure 2. The possibility of decreasing the size and the growth rate (relative to those firms with less than 100 employees).

Second, we can imagine the case where, in actuality, the threshold effect does not exist. This means that most SMEs choose to grow into LSEs without intentional retardation; thus, they give up the benefits that exist for SMEs and accept the regulations for LSEs. This strategic decision can be justifiable if the marginal profit from additional employment is larger than the marginal loss caused by the deprivation of government support and the imposition of regulations. According to Kim's (2005a) research, the proportion of the policy fund for SMEs given to medium sized SMEs with employees from 50 to 299 is only 28.5% in 2005 and 31.8% in 2003. It implies that more than 2/3 of the policy fund for SMEs is given to very small sized firms. The relatively larger firm in medium sized enterprises shows a lower dependency on the policy fund. Moreover, as explained Section 2.2, most regulations on LSEs are imposed to very large firms whose asset size is more than 2 billion dollars. In summary, the firms in the vicinity of less than 300 employees may be already out of policy care and thus, the designation as an LSE does not cause immediate regulation. Therefore, the firms in the vicinity of less than 300 employees have little incentive to retard their growth by giving up the benefit from scale economy.

7. Concluding Remarks

The purpose of this research was to verify whether the policy support for SMEs retards the growth of SMEs. Previous reports and articles insisted on the existence of a threshold effect in the vicinity of less than 300 employees on the basis of surveys or specific cases. However, in that there had been no econometrical empirical research using a database of firms, this research has significance.

The verification of a threshold effect was verified from two perspectives – the possibility of decreasing the size and the growth rate. The regression analysis with data of the total statutory audit firms showed that there was no abnormal growth pattern in the vicinity of less than 300 employees. One possible explanation was the sample bias which was caused by the exclusion of small firms. However, if we considered the fact that most policy supports for SMEs were distributed to very small firms, and regulations for LSEs were imposed on very large firms whose asset value was more than 2 billion dollars, the immediate burden caused by the designation of LSEs was not more than the marginal profit from a scale economy. Therefore, the threshold effect of a firm's growth was not observed in this framework on the assumption of an insignificant small sample bias.

This result is expected to spur more elaborate and improved research coming from the academic field rather than to draw complete conclusion. More serious debate is required to reveal the efficacy of the dichotomous approach between SMEs and LSEs in industrial policy.

REFERENCE

Acemoglu, D. and J.D. Angrist. (2001). "Consequences of employment protection? The case of Americans with Disabilities Act." Journal of Political Economy, Vol 109, No. 5. 915-957.

Ahn, S.H. (2005). "Empirical study on the entry and exit of SMEs', in Kim (eds), The role of SMEs for the innovation-driven economy." KDI Research Report, Korea Development Institute. pp. 210-256.

Borgarello, A., P. Garibaldi and L. Pacelli. (2002). "Employment protection legislation and the size of firms." discussion paper.

Chen, A., Y. Hwang, and B. Shao. (2003). "Measurement and sources of overall and input inefficiencies: Evidences and implication in hospital services." European Journal of Operational Research, Vol. 161, No. 2, 447-468.

Denkabe, P. (2003). "Policy, aid and growth: A threshold hypothesis." discussion paper.

Economist. (2005). "The dilemma of the 300 employees." Economist Oct. 2005.

Kim, H.W. (2005a). "Research on effect of policy support on SME." The 2006 conference of Korean Economic Association.

Kim, J.H., (2005b). "The role of SMEs for the innovation-driven economy." KDI Research Report, Korea Development Institute.

Lee, B. (2003). "The survival of new firms in Korean manufacturing: Start-up conditions and post-entry evolution." Korea Economic Research Institute, Research report 02-26.

National Assembly Budget Office. (2006). Settlement of account of government annual revenue and expenditure, National Assembly Budget Office.

OECD. (2005). Science, technology and industry scoreboard, Paris: OECD.

Schivardi, S. (2004). "Threshold effects and firm size: The case of firing costs." discussion paper.

Son, W.I. (2002). "Corporate income tax in Korea." Korea Institute of Public Finance.

Verick, S. (2004). "Threshold effects of dismissal protection legislation in Germany." discussion paper.

The Federation of Korean Industries. (2006). "Policy agenda of facilitation of SME growth into LSE." The Federation of Korean Industries.

In: Micro-Evidence for the Dynamics of Industrial Evolution ISBN: 978-1-60456-388-7
Editors: A. Heshmati, J.-D. Lee

Chapter 9

EVALUATING DOMINANCE RANKING OF PRODUCTIVITY DISTRIBUTION IN KOREAN MANUFACTURING INDUSTRY

Youngkyu Kim [*]
Technology Management, Economics and Policy Program
College of Engineering, Seoul National University, Seoul, Korea
Inha Oh [†]
Technology Management, Economics and Policy Program
College of Engineering, Seoul National University, Seoul, Korea
Almas Heshmati [‡]
University of Kurdistan Hawler, and
Hawler Institute for Economic and Policy Research
Federal Region of Kurdistan, Hawler, Iraq

ABSTRACT

The competitive entry and exit process of firms allows capital and labor to be reallocated from declining industries to growing industries. It also makes industry and the economy more dynamic. The main purpose of this paper is to analyze whether the effect of productivity on the entry and exit process is positive or negative in the Korean manufacturing industry. The test for first and second order stochastic dominance of total factor productivity level between exit groups and entry or incumbent groups is conducted. The empirical analysis is based on unbalanced plant level panel data obtained from the Annual Report on Mining and Manufacturing Survey in Korea for the period 1993 to 2003. Results show evidence of the negative effect of productivity on the entry and exit process which differs over the years, across different firm sizes and industrial

[*] Corresponding author: Youngkyu Kim, ifrit99@teramail.com
[†] Email: whistle3@snu.ac.kr, inha.oh@gmail.com
[‡] E-mail: heshmati@snu.ac.kr

sectors. After the financial crisis in mid-1997 and 1998, the exit group with less than 50 employees second order dominated the entry group in 1999. A similar result was observed for plants with less than 20 employees in 2002. Our result also shows that the exit group second order dominated the incumbent group in every year when the plants had less than 300 employees. A negative exit effect was also observed frequently in the case of small and medium plants with less than 300 employees, in the post-crisis period and in the case of low-tech firms.

Keywords: Firm dynamics, TFP, Stochastic dominance, Korean manufacturing, SME
JEL Classification Numbers: C40, D24, L22, N65

1. Introduction

The competitive entry and exit process of firms allows capital and labor to be reallocated from declining industries to growing industries. A number of studies including OECD (2003), Stokey (1988), Grossman and Helpman (1991) and Ahn (2001) conducted various theoretical and empirical analyses, regarding the entry and exit process of firms and followed resource reallocations. Generally, it has been said that entry and exit processes which are related to the productivity growth of the entire economy involved processes such as market selection, learning effect and shadow of death. Here market selection means productivity growth effect from the survival of efficient businesses and the exit of inefficient businesses; learning effect means productivity growth of new entrants into the market as time elapses and shadow of death means appearance of lower productivity from exiting firms in advance of their exit.

With regard to the issues above, Hahn (2003) examined the process of creative destruction in the Korean manufacturing industry during 1990-1998 with plant level data. He showed that market selection, learning effect and shadow of death were also observed in the Korean manufacturing industry. And he reported that the processes of firm dynamics, as entry of firms with higher productivity and exit of firms with lower productivity, contributed to the productivity growth as much as 45 to 65 percent in manufacturing industry. The Korean manufacturing industry has been a major source of economic growth from the early 1990s. Since the 1980s, more than 90% of exports from Korea have come from manufacturing industry. (Kim, 2005; Oh et al., 2006).

The main purpose of this paper is to investigate whether the effect of firms' entry and exit on productivity is positive or negative in the Korean manufacturing industry. A positive effect would come from either the survival of more productive firms or the death of less productive firms than average firms. Empirically there are indications of the presence of negative effects. A continuous survival of firms with lower productivity and the death of firms with higher productivity in the industry will have a negative effect on overall productivity. Ahn (2001) analyzed the influence of firms' survival and exit on productivity growth and he presented some results that indicated a negative effect on the productivity growth for the cases of the U.S. (1987 and 1991) and the U.K. (from 1980 to 1992). An OECD (2003) report showed some cases that indicate negative effects arising from the entry and exit of the manufacturing firms on the productivity growth in some countries[1]. It reflected that the survival of firms with lower productivity and the exit of firms with higher productivity is often observed in some cases.

Previous research suggests that firms with lower productivity are facing problems of survival in a competitive market. Also a high debt rate, the exit of sub-coordinating firms and changes in the exchange rate can increase the hazard rate of firms. Productivity also plays a significant role as sources of exit and entry of the firms (Ahn, 2001; Disney et al., 2003; Fukao et al., 2005; Hahn, 2000 and 2003). But Aw et al. (2001)[2] and Foster et al. (2001)[3] showed that the productivity of new entry firms was lower than the productivity of incumbent firms. Firms that entered the market with lower productivity would disappear in a few years or survive in the market by learning effect.

To examine the relationship between productivity and the exit and entry processes in Korean manufacturing industry, we compare the total factor productivity (TFP) level of exit plants with entrants and incumbents based on the Annual Report on Mining and Manufacturing Survey, provided by the Korean national statistical office, for each year from 1993 to 2003. In the survey data, the entry and exit of plants are identified based on the plants appearing and disappearing in the data over time. Entry and exit of plants due to spin-off, split, merger, and acquisition could not be distinguished from ordinary turnover of plants with the available plant level data base. A plant is the most disaggregated unit of significant production and employment in manufacturing industry and plant turnover is an important source of resource reallocation, from the perspective of the national economy. Plant turnover makes a significant contribution to productivity growth, as more productive entering plants replace exiting plants that are less productive. Regardless of the management decisions inside a firm, the opening and closing of a plant should be very costly activities for firms. So, observation of plant level dynamics might be as important as firm level dynamics, and in this research vein. Our approach is in line with a number of empirical studies that focused on plant turnover and related TFP dynamics recently (Aw et al., 2003; Baldwin and Gu, 2006; Ahn et al., 2005; Ahn, 2001; Hahn, 2000; Fukao et al., 2006).

According to the Manufacturing Survey, maintained by the Korean national statistical office, for 1980~2003, the proportion of large plants (with 300 or more employees) in terms of the share of plants has steadily decreased from 3.3% in 1980 to 0.6% in 2003, and those of mid-size (with 50 to 299 employees) and small (with 20 to 49 employees) have also been steadily declining since the late 1980s. The proportion of micro plants (with less than 20 employees), on the other hand, stayed at 57~60% until the mid-1980s, but has rapidly increased since the late 1980s, registering 75.9% in 2003, which has caused a sharp decrease in the average number of employees per plant in the manufacturing sector (Kim, 2005). Bearing these factors in mind, we focused on the TFP level of exiting plants compared to entrants and incumbents in Korean manufacturing industry, considering the size classification of manufacturing industry from 1993 to 2003.

From mid-1997 to 1998, the Korean economy faced the Asian financial crisis and it suffered a sharp decline. In mid-1997, the IMF approved a $58.4 billion standby loan

---

[1] The negative exit effects of firms was observed in the case of the Netherlands (from 1988 to 1993) and in the case of western Germany (from 1992 to 1997)

[2] Aw et al. (2001) observed lower productivity of the new firms in the case of the Taiwan manufacturing industry in 1986.

[3] Foster et al. (2001) showed the lower productivity in the case of American manufacturing industry.

arrangement to build foreign exchange reserves. This rescue plan included a range of structural reforms in the financial and corporate sectors to address what the IMF believed to be the causes of the crisis. The government approved the IMF reform package, and a tight monetary policy followed immediately. As a result, the overnight call rate shot up to 25 percent. The currency market turbulence quickly crushed the banking sector. By early 1998, most commercial banks and other financial institutions were in technical default due to severe depreciation and high interest rates. Damage to the economy from the twin crises and the tight monetary policy was substantial. Real GDP declined by 8.1 percent in the third quarter of 1998 compared with the previous pre-crisis year. (Koo and Kiser, 2001)

The unexpected negative and sharp decline in the Korean economy led to serious efforts to be made to use the existing potential to achieve a rapid recovery. The Korean economy recovered rapidly from the deep crisis and it grew rapidly, by an average of 7.25% per annum from 1999 to 2002. In this period, the government determined to increase employment and technology innovation by fostering small and medium sized enterprises (SMEs). The government intervened strongly in the entire process of certification, supporting and financing of the venture business. (Oh et al., 2006)

In this study, we conducted tests of the first and the second order stochastic dominance between two distributions of TFPs. One is the cumulative distribution function TFP level of exiting plants, and the other is the cumulative distribution function of entrant and incumbent plants TFP level. Two statistical tests for detecting stochastic dominance in distribution have been proposed recently by Barrett et al. (2003) and Linton et al. (2004). This test was based on the Kolmogorov-Smirnov (K-S) test, and compared the two candidate cumulative distribution functions at all points in the sample. The Kolmogorov-Smirnov test is usually used in statistics literature to compare sample distributions. To test for the statistical significance of stochastic dominance we used the approaches suggested in Maasoumi and Heshmati (2005) and Linton et al. (2004). In those studies, it was used to test for differences of income distribution of individuals and households.

Using the approach above, this paper compares the TFP level of each group by utilizing the stochastic dominance scheme. And as far as we know, this is the first attempt to apply the stochastic dominance methodology to analyze productivity differences in manufacturing. Other research which aimed to compare the TFP of entry and exit plants calculated the statistical significance of difference using t-statistics. For larger datasets, the central limit theorem suggested that the t-test will produce valid results even in the face of non-normally distributed data. However, highly non-normal datasets can cause the t-test to produce false results, even for large datasets. In this study we produce results that confirm the difference of two statistical tests. The compared empirical result of mean difference with statistical significance using t-test and test of stochastic dominance suggest that the two TFP distributions are not the same at the 5% of significant level.

This paper also analyzed the plant level data of manufacturing industry from 1993 to 2003, with firm size and industrial sector classifications. The analysis period allowed us to compare the results before the finance crisis period and post-crisis period in Korea, which occurred from mid-1997 to 1998.

The rest of this study is organized as follows. In the next section the data is described. The stochastic dominance methodology is outlined in Section 3. The discussion of the results is found in Section 4, followed by the summary of results in Section 5.

2. DATA

We use a sample data that contains a significant number of entry and exit plants. The survey data used is unbalanced plant level panel data based on the Annual Report on Mining and Manufacturing Survey, provided by the Korean national statistical office, for each year from 1993 to 2003. The survey data consist of 310,866 distinct plants in 580 manufacturing industries at KSIC (Korean Standard Industrial Classification) five-digit level with five or more employees. The number of observations on the survived plant both in the survey and in the panel data is the same. The total number of observations was 1,030,846. In the survey data, the entry and exit of plants were identified based on the plants' appearance and disappearance over time. Entry and exit of plants due to spin-off, split, merger, and acquisition could not be distinguished from ordinary turnover of plants with the available plant level data base.

The panel data include industrial classification number, number of employees, TFP level, first year observed, last year observed and identity number of observed plants during their surviving years. If the identity number of observations is the same, these observations are considered to come from the same plant. When a plant appears or disappears over time, first year and last year of observation of the plant are used as the entry time and exit time, in the panel data. We eliminated the 1993 entering year and 2003 exit year rows because we do not know the entry and exit dates of these plants.

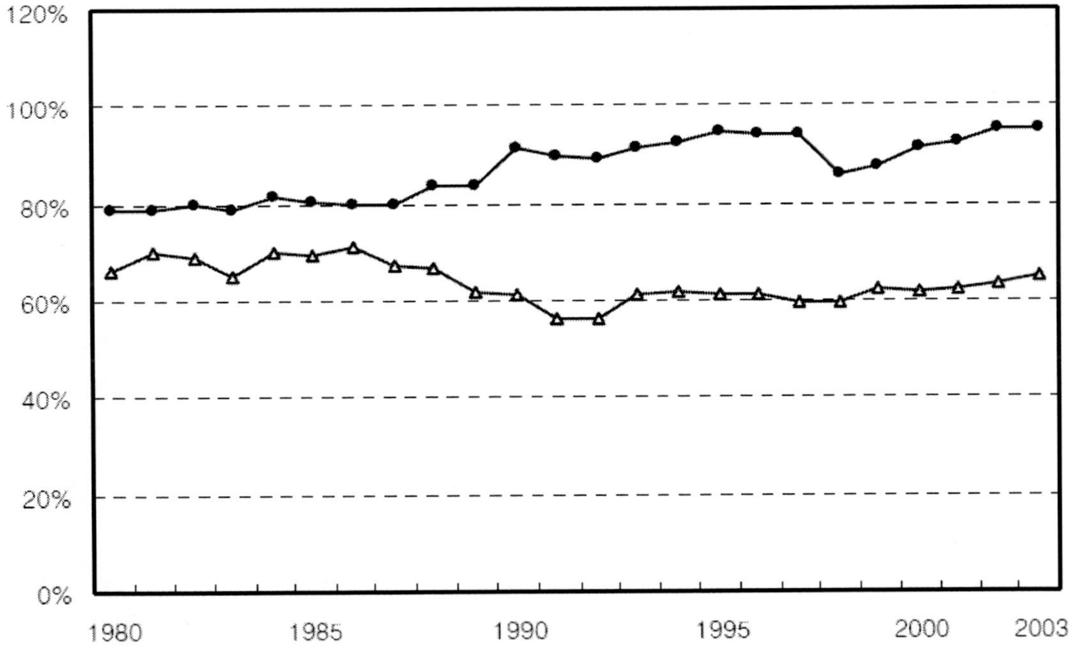

Note: The upper line is output production weight of mining and manufacturing survey data over whole manufacturing data; The bottom line is employee weight of mining and manufacturing survey data over whole manufacturing data.

Figure 1. A Representative of Mining and Manufacturing Survey data (Korea Development Institute (KDI)).

We calculated total factor productivity (TFP) of each plant using a chained-multilateral index developed in Good (1985) and applied by Aw et al. (2001). The methodology adopted is fully explained in Appendix A. The TFP level was transformed into natural logarithms of TFP to transform it into a normal distribution

To compare the TFP level of entry, exit and incumbent plants with classification by size, we separated plants by employees each year: (1) Plants with below 20 employees (754,902 observations), (2) Plants with between 20 and 49 employees (180,526 observations), (3) Plants with between 50 and 299 employees (83,088 observations), and (4) Plants with more than 300 employees (7,830 observations).

The Annual Report on Mining and Manufacturing Survey does not provide data for all plants in the manufacturing industry. Because the survey data include plants that have five or more employees. Many plants with below 5 employees are eliminated from the survey data in manufacturing industry (Kim, 2005). Figure 1 shows the production weight and employee weight of all manufacturing industry. The production weight of the survey data plants corresponds to about 80% until the mid-1980s and it exceeded 90% after the early 1990s. But the employee weight of the survey data plants corresponds to about 70% until the mid-1980s but it declined to about 60% after the mid-1980s (Kim, 2005).

It should be noted that the plant-level turnover of these datasets not only reflects entries and exits of plants, but also includes plants crossing the survey threshold (five or more employees). The share of plants with employees of less than five in the total number of plants was 63%, but their share accounted for only 12% of total employment, 3% in gross sales and 4% in gross value added of the whole manufacturing industry, according to the quinquennial Industrial Census in 2003. Therefore we assumed that biases associated with the exclusion of these micro plants might be relatively small. Additionally, it was observed that the exit rate of micro plants was very high, and such fragile micro plants may not be good representative samples (Oh et al., 2006).

We also grouped the plants by using the industrial classification employed in the OECD (2003) industrial classification[4] into: (1) Plants with high technology, (2) Plants with medium high technology (3), Plants with medium low technology, and (4) Plants with low technology.

3. METHODOLOGY

Testing for stochastic dominance has generally been used to compare income distributions, although in recent years, the methodology has been used in finance to test dominance in returns distributions. Maasoumi and Heshmati (2005) evaluated dominance ranking of PSID (panel study of income dynamics) incomes by this methodology. This

---

[4] *High technology* plants include: Pharmaceuticals, Office accounting and computing machinery, Radio television and communication equipment, Aircraft and spacecraft.

Medium High technology plants include: Chemicals excluding pharmaceuticals, Machinery and equipment n.e.c., Electrical machinery and apparatus n.e.c., Medical precision and optical instruments, Motor vehicles trailers and semi-trailers Railroad equipment and transport equipment n.e.c.

Medium Low technology plants are: Coke refined petroleum products and nuclear fuel, Rubber and plastics products, Other non-metallic mineral products, Basic metals, Fabricated metal products except machinery and equipment, Building and repairing of ships and boats, Manufacturing n.e.c.; recycling

Low technology plants consist of: Food products beverages and tobacco, Textiles textile products leather and footwear, Wood and products of wood and cork, Pulp paper, paper products printing and publishing.

methodology tries to determine if two datasets differ statistically in distributions. This test has the advantage of not making a priori any assumptions about the distribution of data.

In this study, instead of income distributions, we compared two cumulative distributions of TFP levels using the first and second orders of stochastic dominance. To test for stochastic dominance we use the same approach as Maasoumi and Heshmati (2005). This approach is similar to the Linton et al. (2004) approach. However there are some differences between the two studies. Maasoumi and Heshmati (2005) did not impose the boundary of the null of dominance and obtained unconstrained estimates of the probabilities of non-rejection in the actual samples. This allowed a classical hypothesis testing as confidence intervals that avoided the null hypothesis bias of the frequent method.

The tests performed in Maasoumi and Heshmati (2005) compared the cumulative distribution functions at all points in the sample distribution and was based on the Kolmogorov-Smirnov (K-S) test. K-S test is used in the statistics literature to compare two sample distributions. The K-S test has the advantage of making no assumptions about the distribution of data. For larger datasets, the central limit theorem suggested that the t-test will produce valid results even in the face of non-normally distributed data. However, highly non-normal datasets can cause the t-test to produce false results, even for large N datasets. For example, consider the following two data sets.

Set A = {1.26, 0.34, 0.70, 1.75, 50.57, 1.55, 0.08, 0.42, 0.50, 3.20, 0.15, 0.49, 0.95, 0.24, 1.37, 6.98, 0.1, 0.94, 0.38} and
Set B = {2.37, 2.16, 14.82, 1.73, 41.04, 0.23, 1.32, 2.91, 39.41, 0.11, 27.44, 4.51, 0.51, 4.5, 0.18, 14.68, 4.66, 1.3, 2.06, 1.19}

The average of set A = 3.6125, while the *average of set B* = 8.3565, *The p-value (T-test for mean-difference)* = 0.2198 and the *D-statistic (K-S test for two datasets)* = 0.45.

These datasets were drawn from lognormal distributions that differ substantially in mean. The K-S test neglects the level differences but detects the difference in the distribution, while the t-test accounts for the level difference but it does not detect the difference in distribution. Of course, if the researcher knew that the data were non-normally distributed, they would not apply the t-test in the first place.

In introducing the stochastic dominance methodology we follow the definitions proposed and notations used by Maasoumi and Heshmati (2005). Let X_1 and X_2 be two variables at either two different points in time, or for different regions or countries, or with or without a program (treatment). Let X_{ki}, i = 1, ..., N; k = 1, 2 denote the not necessarily i.i.d. observations.

Let U_1 denote the class of all von Neumann-Morgenstern type utility functions, u, such that $u' \geq 0$, (increasing). Also, let U_2 denote the class of all utility functions in U_1 for which $u'' \geq 0$ (strict concavity), and U_3 denote a subset of U_2 for which $u''' \geq 0$. Let $X_{(1p)}$ and $X_{(2p)}$ denote the p-th quantiles, and $F_1(x)$ and $F_2(x)$ denote the cumulative distribution functions, respectively.

Definition: X_1 First Order Stochastic Dominates X_2, denoted X_1 FSD X_2, if any of the following equivalent conditions holds:

$E[u(X_1)] \geq E[u(X_2)]$ for all $u \in U_1$ with strict inequality for some u'; or (1)

$F_1(x) \leq F_2(x)$ for all x with strict inequality for some x ; or (2)

$X_{(1p)} \geq X_{(2p)}$ for all $0 \leq p \leq 1$, with strict inequality for some p. (3)

Definition: X_1 Second Order Stochastic Dominates X_2, denoted X_1 SSD X_2, if any of the following equivalent conditions holds:

$E[u(X_1)] \leq E[u(X_2)]$ for all $u \in U_2$, with strict inequality for some u ; or (4)

$\int_{-\infty}^{x} F_1(t)dt \leq \int_{-\infty}^{x} F_2(t)dt$ for all x with strict inequality for some x; or (5)

$\Phi_1(p) = \int_0^p X_{(1t)}dt \geq \Phi_2(p) = \int_0^p X_{(2t)}dt$ for all $0 \leq p \leq 1$, with strict inequality for some value(s) p.

(6)

Maasoumi and Heshmati (2005) proposed the following cumulative distribution function and test statistics: For k=1,2 define: for θ

$$F_k(x,\theta) = P(X_{ki}(\theta) \leq x) \tag{7}$$

and

$$\overline{F}_{kN}(x,\theta) = \frac{1}{N}\sum_{i=1}^{N} 1(X_{ki}(\theta) \leq x). \tag{8}$$

The authors denote $F_k(x) = F_k(x,\theta_{k0})$ and $\overline{F}_{kN}(x) = \overline{F}_{kN}(x,\theta_{k0})$, and let $F(x_1,\ldots\ldots,x_k)$ to be the joint c.d.f. of $(X_1,X_2)'$. They define the following functions of the joint distribution:

$$d = \min_{k \neq l} \sup_{x \in X} [F_k(x) - F_l(x)] \tag{9}$$

$$s = \min_{k \neq l} \sup_{x \in X} \int_{-\infty}^{x} [F_k(t) - F_l(t)]dt \tag{10}$$

where X denotes a given set contained in the union of the supports of X_{ki} for k=1,2. Without loss of generality they assume that the supports are bounded. The hypotheses of interest can now be stated as:

$$H_0^d : d \leq 0 \ vs. \ H_1^d : d \succ 0 \tag{11}$$

$$H_0^s : s \leq 0 \ vs. \ H_1^s : s \succ 0 \tag{12}$$

The test statistics they considered are based on the empirical analogues of (9)-(10). They are defined to be:

$$D_N = \min_{k \neq l} \sup_{x \in X} \sqrt{N}[\overline{F}_{kN}(x, \dot{\theta}_k) - \overline{F}_{lN}(x, \dot{\theta}_l)] \tag{13}$$

$$S_N = \min_{k \neq l} \sup_{x \in X} \sqrt{N} \int_{-\infty}^{x} [\overline{F}_{kN}(t, \dot{\theta}_k) - \overline{F}_{lN}(t, \dot{\theta}_l)] dt \tag{14}$$

In their applications they reported Probability { $D_N \leq 0$ } and Probability { $S_N \leq 0$ } and were able to identify which distribution dominates, if any. They also reported the critical level associated with this non-rejection region. In this study we employ the above approach to analyze the distribution of total factor productivity growth in Korean manufacturing.

4. Results and Discussions

We now turn to the results of the statistical tests obtained from the stochastic dominance test procedures, which are: (1) between TFP levels of entry plants and exit plants from 1994 to 2002, and (2) between TFP levels of incumbent plants and exit plants every year from 1994 to 2002. In our empirical analysis we focused on results based on the tests for stochastic dominance outlined in Maasoumi and Heshmati (2005). We set the interval of distribution to 5%, and bootstrapped the TFP level samples 100 times for each distribution. We set the statistical significance level of stochastic dominance as more than 80% probability level following Maasoumi and Heshmati (2005).

If exit group first order dominates the entry group, the cumulative function value of exit plants TFP level is lower than value of entry plants in all distribution areas. It means that an obvious significant difference between two groups is observed. We tested stochastic dominance with second order when we did not observe any first order dominance. In the case of second order dominance, we cumulated once again the function which was cumulated for first order dominance. It also provides the difference between two groups stochastically. Maasoumi and Heshmati (2005) showed the second order dominance to compare two groups.

We analyzed whether the effect of entering and exiting on the productivity process is positive or negative using stochastic dominance relation. A positive effect comes from the survival of higher productivity plants and the death of lower productivity plants. But the reserve situation or continuous surviving of lower productivity plants and death of higher productivity plants in the industry has a negative effect on productivity. If a productivity distribution of the exit group stochastically dominates a productivity distribution of the incumbent group, it shows negative effect on productivity of the exit group. Such information can be used in decisions of allocation of pubic support. Public support has been a vital instrument affecting the distribution of firms in manufacturing with great impact on the economic growth and dynamic structure of the industry.

4.1. Test Result Comparing Entrant and Exit Plants

The first parts of our tests followed a two-step procedure: (1) we tested whether the TFP level of exit plants stochastically dominates the TFP level of entry plants for different size classes, and (2) we conducted the same procedure with industrial classification in each size class.

In Table 1, the results comparing entrants and exit plants by size classification are summarized. There was no first order stochastic dominance between entrants and exit plants. In comparing plants exiting the market, the TFP level with entrant plants, in total 3 cases (1996, 1999 and 2002) of second order dominance are observed and there was no dominance in the other cases. The TFP level of entering plants dominates the TFP level of exit plants in 1996. In 1999 and 2002, the exit group second order dominates the entrant group.

Table 1. The Stochastic Dominance Relation between Entry Plants and Exit Plants

	1994	1995	1996	1997	1998	1999	2000	2001	2002
Total									
First-Order									
Second-Order			En			Ex			Ex
Below 20									
First-Order									
Second-Order			En			Ex			Ex
From 20 to 49									
First-Order									
Second-Order				En		Ex	En		
From 50 to 300									
First-Order									
Second-Order	En			En			En	Ex	En
More than 300									
First-Order									
Second-Order									

Note: En means TFP level of entry stochastic dominates TFP level of exit; Ex means TFP level of exit stochastic dominates TFP level of entry.

Table 2. The Stochastic Dominance Relation between Incumbent Plants and Exit Plants

	1994	1995	1996	1997	1998	1999	2000	2001	2002
Total									
First-Order									
Second-Order	Ex	Ex	Ex	Ex	Ex	Ex	Ex	Ex	Ex
Below 20									
First-Order									
Second-Order	Ex	Ex	Ex	Ex	Ex	Ex	Ex	Ex	Ex
From 20 to 49									
First-Order									
Second-Order	Ex	Ex	Ex	Ex	Ex	Ex	Ex	Ex	Ex
From 50 to 300									
First-Order									
Second-Order	Ex	Ex	Ex	Ex	Ex	Ex	Ex	Ex	Ex
More than 300									
First-Order									
Second-Order		Ex							

Note: En means TFP level of entry stochastic dominates TFP level of exit; Ex means TFP level of exit stochastic dominates TFP level of entry.

In the case of plants with less than 20 employees, there were 3 cases (1996, 1999 and 2002), where second order dominance was observed. These results were the same as the tests for total sample plants, because the number of observations included in the size group of below 20 employees was 73% of total observations. In the case of the size group categorized as 20-49 employees, we observed second order dominance in 1997, 1999 and 2000. Only in 1999 years, the exit group second order dominated the entry group. There were 5 observations (1994, 1997, 2000, 2001 and 2002) in the case of plant size group with 50-300 employees; Entry group second order dominated the exit group in four out of five cases except for 2001. For firms with more than 300 employees, there is no dominance observed between entry and exit groups.

After the financial crisis in mid-1997 and 1998, here called the post-crisis period, the exit group second order dominated the entry group for the first time in 1999 (in the case of plants with less than 50 employees). It means that the exit group had relatively much higher TFP level plants than the entry group in 1999 (in the case of plants with less than 50 employees). In other words, many lower TFP level plants with less than 50 employees entered the market relative to exit plants in 1999. The entry rates stayed above 20% most of the time except during the financial crisis (1997-1998) period and the entry rates peaked at 31.1% in 1999 (Kim, 2005). Table 3 and Figure 2 report the number of entry and exit plants by size classification in every year. After the financial crisis, the number of entry plants increased to about 72% in 1999 from the previous year in the case of plants with less then 20 employees, about 43% in the case of plants with 20-49 employees and about 52% in the case of plants with 50-300 employees. However, the exit group second order dominates the entry group in 1999 in the case of plants with less than 50 employees but not of plants with 50-300 employees. It seems that lower TFP level plants with less than 50 employees entered the market more frequently than plants with more than 50 employees in 1999.

Table 3. The Number of Entry Plants and Exit plants for different size classes

	Below 20		From 20 to 49		From 50 to 300	
	Entry	Exit	Entry	Exit	Entry	Exit
1994	17103	16645	3012	3090	895	978
1995	21985	15314	3007	2451	733	774
1996	16578	18046	2261	2466	557	712
1997	14090	19796	1726	2551	502	758
1998	11856	12767	1680	1719	369	437
1999	20476	13193	2417	1773	561	463
2000	20218	17607	2574	2363	627	759
2001	22564	27020	2949	2820	811	703
2002	30807	16828	3289	1809	815	553

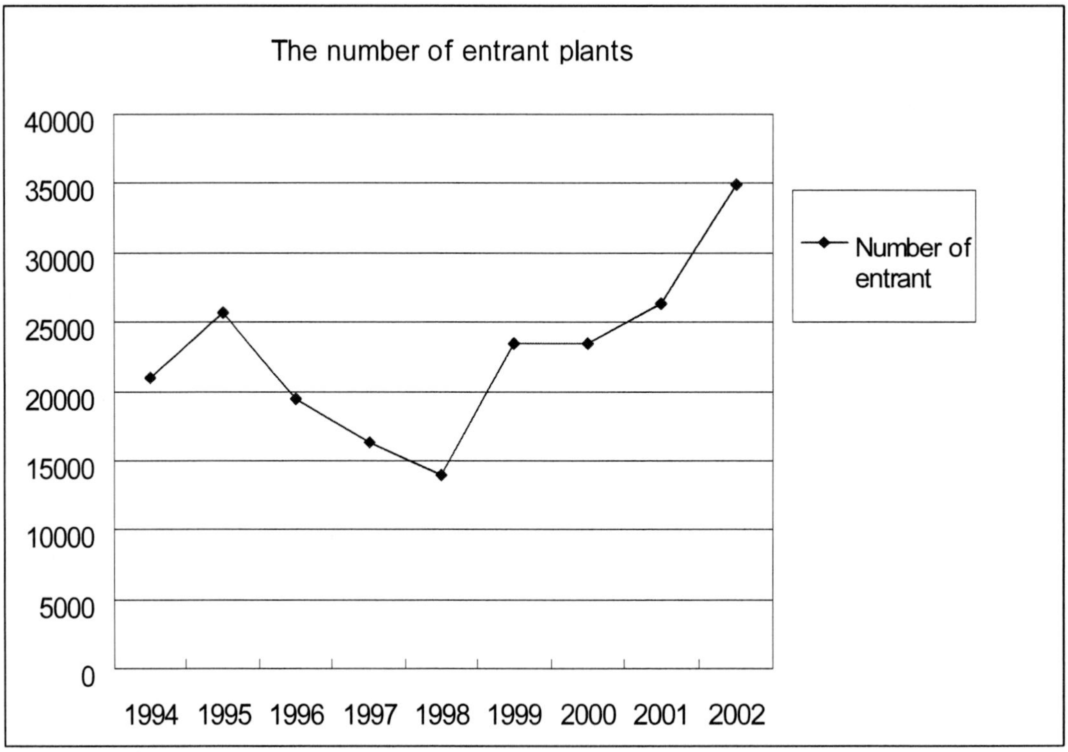

Figure 2. A Number of Entrant Plants with Bellow 300 Employees.

A similar case is observed for plants with less then 20 employees in 2002. In Table 3, the number of entry plants with less then 20 employees increased about 36% compared with the previous year in 2002 and about 11.5% in the case of plants with 20-49 employees and about 0% in the case of plants with 50-300 employees. We observed second order dominance in 2002 for plants with less then 20 employees. We also observed a negative effect in the case of plants with 50-300 employees in 2001.

These negative effects of productivity were also different by industrial classification reported in Tables 4 to 6. There was only one negative effect of productivity in the case of medium high technology (1996) in Table 6 before the financial crisis period. But it could be included in the financial crisis period because 1996 year means that the disappearance year was 1997. Especially, we found negative effects in the post-crisis period for high technology and medium high technology cases (20-300), and for medium low technology and low technology cases of less than 20 employees.

Table 4. The Stochastic Dominance Relation between Entry Plants and Exit Plants as Industry Classification bellow 20 employees

	1994	1995	1996	1997	1998	1999	2000	2001	2002
High technology									
First-Order									
Second-Order		En		En					
Medium High technology									
First-Order									
Second-Order				Ex					
Medium Low technology									
First-Order									
Second-Order	En		En		En	Ex			
Low technology									
First-Order									
Second-Order				Ex					Ex

Note: En means TFP level of entry stochastic dominates TFP level of exit; Ex means TFP level of exit stochastic dominates TFP level of entry.

Table 5. The Stochastic Dominance Relation between Entry Plants and Exit Plants as Industry Classification from 20 to 50 employees

	1994	1995	1996	1997	1998	1999	2000	2001	2002
High technology									
First-Order									
Second-Order						Ex			Ex
Medium, High technology									
First-Order									
Second-Order		En					En	Ex	
Medium, Low technology									
First-Order									
Second-Order	En			En					
Low technology									
First-Order									
Second-Order	Ex				Ex	Ex			

Note: En means TFP level of entry stochastic dominates TFP level of exit; Ex means TFP level of exit stochastic dominates TFP level of entry.

Table 6. The Stochastic Dominance Relation between Entry Plants and Exit Plants as Industry Classification from 50 to 300 employees

	1994	1995	1996	1997	1998	1999	2000	2001	2002
High technology									
First-Order									
Second-Order	En					Ex			Ex
Medium, High technology									
First-Order									
Second-Order	En		Ex					Ex	
Medium, Low technology									
First-Order									
Second-Order	En		En	En	En	Ex	En		
Low technology									
First-Order									
Second-Order				Ex					

Note: En means TFP level of entry stochastic dominates TFP level of exit; Ex means TFP level of exit stochastic dominates TFP level of entry.

4.2. Test Result Comparing Incumbents and Exit Plants

The second parts of our tests included a two-step procedure: (1) we tested whether the TFP level of exit plants stochastically dominates the TFP level of incumbent plants for different size classes, and (2) we conducted the same tests but for firms of different sizes operating in different industrial sectors.

The results are summarized in Table 2. In this testing procedure, we compared the exit group with the incumbent group. In the case of total plants, exit firms second order dominates incumbents in every year. There exists second order dominance in every year and size classification with the exception of plants with more than 300 employees. In the case of plants with more than 300 employees, exit plants second order dominates the incumbent firms but only in 1995.

These results show the negative exit effect of productivity in the Korean manufacturing industry. It means that incumbent groups have more low TFP level plants relative to the exit group in every year. It means also that lower TFP level plants with less than 300 employees have appeared in the incumbent group without disappearing from the market. In industrial classification, these negative effects were observed frequently among low technology industry and in the post-crisis period reported in Table 7-9. In conclusion, we could observe negative exit effects frequently in the case of small and medium plants (below 300 employees), in the case of the post-crisis period and in the case of low technology industries.

Table 7. The Stochastic Dominance Relation between Incumbent Plants and Exit Plants as Industry Classification bellow 20 employees

	1994	1995	1996	1997	1998	1999	2000	2001	2002
High technology									
First-Order									
Second-Order						Ex		Ex	Ex
Medium, High technology									
First-Order									
Second-Order	Ex	Ex	Ex	Ex	Ex	Ex	Ex	Ex	Ex
Medium, Low technology									
First-Order									
Second-Order	Ex	Ex	Ex	Ex	En	Ex	Ex	Ex	Ex
Low technology									
First-Order									
Second-Order	Ex	Ex	Ex	Ex	Ex	Ex	Ex	Ex	Ex

Note: En means TFP level of entry stochastic dominates TFP level of exit; Ex means TFP level of exit stochastic dominates TFP level of entry.

Table 8. The Stochastic Dominance Relation between Incumbent Plants and Exit Plants as Industry Classification from 20 to 50 employees

	1994	1995	1996	1997	1998	1999	2000	2001	2002
High technology									
First-Order									
Second-Order	Ex			Ex	Ex	Ex		Ex	Ex
Medium, High technology									
First-Order									
Second-Order			Ex		Ex	Ex	Ex	Ex	Ex
Medium, Low technology									
First-Order									
Second-Order	Ex	Ex	Ex	Ex	Ex	Ex	Ex	Ex	Ex
Low technology									
First-Order									
Second-Order	Ex	Ex	Ex	Ex	Ex	Ex	Ex	Ex	Ex

Note: En means TFP level of entry stochastic dominates TFP level of exit; Ex means TFP level of exit stochastic dominates TFP level of entry.

Table 9. The Stochastic Dominance Relation between Incumbent Plants and Exit Plants as Industry Classification from 50 to 300 employees

	1994	1995	1996	1997	1998	1999	2000	2001	2002
High technology									
First-Order									
Second-Order		Ex		Ex	Ex	Ex	Ex	Ex	Ex
Medium, High technology									
First-Order									
Second-Order		Ex	Ex	Ex	Ex	Ex	Ex	Ex	Ex
Medium, Low technology									
First-Order									
Second-Order					Ex	Ex	Ex	Ex	Ex
Low technology									
First-Order									
Second-Order	Ex	Ex	Ex	Ex	Ex	Ex	Ex	Ex	Ex

Note: En means TFP level of entry stochastic dominates TFP level of exit; Ex means TFP level of exit stochastic dominates TFP level of entry.

We would like to point out that these negative effects of entry and exit processes are also influenced by other external factors in the post-crisis period. The most feasible external factor is the government policy of financial support to small and medium-sized enterprises (SMEs) such as the public credit guarantee program during the post-crisis period. But government policy has distorted the competitive 'Firm dynamics' process (IMF, 2005). As a result, plants with higher productivity disappeared from the market. Kang (2005) also analyzed the role of the credit guarantee scheme to the SMEs' survival and performance. For negative productivity effect on the exit process, it has been also found that there was negative exit effect on the entire manufacture industry of Japan from 1994 to 2001 in Ahn et al. (2005). They suggested that the existence of zombie firms and relocation of their manufacturing abroad caused the negative exit effect.

5. Summary and Conclusions

The main purpose of this paper was to investigate whether in recent years the effect of firms' entry and exit processes on productivity has been positive or negative in the Korean manufacturing industry. A positive effect would have come from the survival of higher productive firms and the death of lower productive firms. But the continuous survival of lower productive firms and the death of higher productive firms in the industry lead to a negative effect on the overall productivity growth.

Table 10. Significance Test for Mean Difference between Entry plant and Exit plant

	1994	1995	1996	1997	1998	1999	2000	2001	2002
Below 20									
Mean difference	-0.0171	-0.0113	0.0026	-0.0153	0.0083	-0.0247	-0.0101	0.0063	-0.0055
P-value	0.0432	0.1563	0.7522	0.0494	0.4461	0.0032	0.2145	0.3083	0.4042
Dominance	N	N	En	N	N	Ex	N	N	Ex
From 20 to 49									
Mean difference	-0.0260	0.0012	-0.0091	0.0292	0.0152	-0.0338	0.0457	-0.0152	0.0122
P-value	0.2789	0.9644	0.7447	0.3687	0.7695	0.3478	0.3546	0.5677	0.7560
Dominance	N	N	N	En	N	Ex	En	N	N
From 50 to 300									
Mean difference	0.0877	0.0715	-0.0350	0.2268	0.0445	-0.0700	0.2065	-0.1198	0.1540
P-value	0.0179	0.0821	0.5722	0.0150	0.5880	0.3995	0.0186	0.4726	0.1526
Dominance	En	N	N	En	N	N	En	Ex	En
More than 300									
Mean difference	0.1863	-0.1716	-0.3263	0.0147	-0.1143	0.0273	-0.0856	-0.0197	0.0316
P-value	0.6484	0.1558	0.0284	0.9393	0.6180	0.8683	0.4885	0.8669	0.7434
Dominance	N	N	N	N	N	N	N	N	N

Note: Mean difference: (mean of lnTFP in Entry plant – mean of lnTFP in Exit plant).
Dominance means second order stochastic dominance relation between entry plant and exit plant. N means there is no stochastic dominance; En means TFP level of entry stochastic dominates TFP level of exit; Ex means TFP level of exit stochastic dominates TFP level of entry.

We examined two empirical results using tests for the stochastic dominance approach from a firm dynamics perspective in Korean manufacturing industry. Firstly, we investigated whether the entrants have lower productivity than exit plants, i.e. a negative effect. The results suggest that the effect of productivity on entrants is positive in many case studies. The 'Vintage effect' assumption also suggested that entrants have higher productivity than incumbent plants. But there are also studies that show different results. Hahn (2003) showed empirical results about the negative effect of productivity on the entrant process and he analyzed market selection and the learning effect of the entrant plants. Hahn also showed that productivity of entrant plants is even lower than that of exit plants and that means a negative effect of productivity on the entry process of firms. In this paper, we examined stochastic dominance for productivity distribution of two groups between the entry plant group and exit plant group with different sizes and industrial classifications.

Secondly, we conducted tests of the stochastic dominance of productivity distribution between the incumbent plants group and the exit plants group for observing possible negative effects on productivity. The continuous survival of firms with lower productivity and the death of firms with higher productivity in the industry imposes a negative effect on productivity. Ahn (2001) investigated the influence of productivity on firm survival/exit and he presented some results that reported a negative effect in the U.S. (1987 and 1991) and the U.K. (from 1980 to 1992) on productivity. The OECD (2003) report showed evidence in some cases that there was a negative effect of entry and exit firms on productivity growth in some countries. It should be noted that Ahn did not use the stochastic dominance methodology.

We observed the egative effects of productivity on entry and exit processes differently over years, across size classes, and industrial sectors. There is no stochastic dominance among large plants with more than 300 employees except for one case (1995 years) reported. But we observed stochastic dominance in small and medium plants with below 300 employees in several cases. So we paid attention to the case of small and medium plants with less than 300 employees.

Table 11. Significance Test for Mean Difference between Incumbent plant and Exit plant

	1994	1995	1996	1997	1998	1999	2000	2001	2002
Below 20									
Mean difference	-0.0721	-0.0719	-0.0594	-0.0537	-0.1019	-0.0724	-0.0546	-0.0598	-0.0799
P-value	0.0000	0.0000	0.0000	0.0000	0.0000	0.0000	0.0000	0.0000	0.0000
Dominance	Ex	Ex	Ex	Ex	Ex	Ex	Ex	Ex	Ex
From 20 to 49									
Mean difference	-0.0823	-0.1137	-0.0933	-0.0951	-0.2303	-0.1988	-0.1835	-0.1754	-0.2292
P-value	0.0000	0.0000	0.0000	0.0000	0.0000	0.0000	0.0000	0.0000	0.0000
Dominance	Ex	Ex	Ex	Ex	Ex	Ex	Ex	Ex	Ex
From 50 to 300									
Mean difference	-0.0337	-0.0383	-0.0773	-0.0758	-0.2217	-0.2286	-0.2880	-0.6393	-0.4011
P-value	0.1248	0.1076	0.0045	0.0041	0.0000	0.0000	0.0000	0.0000	0.0000
Dominance	Ex	Ex	Ex	Ex	Ex	Ex	Ex	Ex	Ex
More than 300									
Mean difference	-0.1272	-0.1048	-0.2986	-0.0431	-0.1222	-0.0041	0.0601	0.0417	0.0572
P-value	0.1340	0.1553	0.0005	0.5588	0.2010	0.9676	0.4730	0.5713	0.5444
Dominance	N	Ex	N	N	N	N	N	N	N

Note: Mean difference: (mean of lnTFP in Incumbent plant – mean of lnTFP in Exit plant).

Dominance means second order stochastic dominance relation between incumbent plant and exit plant. N means there is no stochastic dominance; I means TFP level of entry stochastic dominates TFP level of exit; Ex means TFP level of exit stochastic dominates TFP level of entry.

Our results also showed that the exit group second order dominates the incumbent group in every year in the case of plants with less than 300 employees in every year.

APPENDIX A. THE MEASUREMENT OF TFP

The Total Factor Productivity (TFP) of each firm or plant can be computed using the chained-multilateral index number approach developed in Good (1985), Good et al. (1997) and applied by Hahn (2000). This method uses a separate reference point for each cross-section of observations and then chain-links the reference points together over time. The reference point for a given time period is constructed as a hypothetical firm or plant with input shares that equal the arithmetic mean input shares and input levels that equal the geometric mean of the inputs over all cross-section observations. Thus, the output, inputs, and productivity level of each firm or plant in each year is measured relative to the hypothetical plant at the base time period. This approach allows us to make transitive comparisons of productivity levels among observations in a panel data.

Specifically, in the context of this study, the productivity index for plant i at time t in our study using the non-parametric approach is measured in the following way:

$$\ln TFP_{it} = (\ln Y_{it} - \overline{\ln Y_t}) + \sum_{\tau=2}^{t} (\overline{\ln Y_\tau} - \overline{\ln Y_{\tau-1}}) -$$
$$\left\{ \sum_{n=1}^{N} \frac{1}{2} (S_{nit} + \overline{S_{nt}})(\ln X_{nit} - \overline{\ln X_{nt}}) + \sum_{\tau=2}^{t} \sum_{n=1}^{N} \frac{1}{2} (\overline{S_{n\tau}} + \overline{S_{n\tau-1}})(\overline{\ln X_{n\tau}} - \overline{\ln X_{n\tau-1}}) \right\}$$

where Y , X , S , and TFP denote output, input, input share, and TFP level, respectively. The symbols with upper bar are corresponding measures for hypothetical plants. The subscripts k and n are indices for time and inputs, respectively.

A number of variables in the survey were used to compute TFP levels. These include input and output variables. The gross production of each plant was used as a measure of output, while capital, labor and materials were used as input variables. The cost shares were obtained as the respective input factors share of the total cost of capital, labor and materials.

An alternative measure to gross output production was the value added. Gross production was used to utilize information on material input levels. Materials are proportional to output, however, variations in material inputs affect the economies of scale. Additionally, the source can be different, representing different advantages or constraints to different industries. Output was deflated by the producer price index at the three-digit industry level.

As a measure of capital stock, the average book value of capital stocks was used at the beginning and end of the year, deflated by the capital goods deflator. The capital in Korean manufacturing has been traditionally used intensively with extremely little loss in the rate of capacity utilization. Thus, the book value of capital stock was found to be a good measure of capital input.

As a measure of labor input, the number of workers was used, this includes paid employees (production and non-production workers), working proprietors and unpaid family workers. Here, the qualitative differential between production workers and all the other types of workers was accommodated. The labor quality index of the latter was calculated as the ratio of non-production workers' and production workers' cumulative wage, divided by the number of workers involved in non-production and production activities for each year.

Finally, as a measure of intermediate input, the "major production costs" plus "other production cost" was used in the survey. Major production costs covered costs arising from materials and parts, fuel, electricity, water, outsourced manufactured goods and maintenance. Additional production costs covered outsourced services, such as advertising, transportation, communication and insurance. The estimated intermediate input was deflated by the intermediate input price index.

When computing TFP growth based on a non-parametric approach, constant returns of scale were assumed. Therefore, the sum of factor input elasticity is equal to one. Labor and intermediate input elasticity for each plant were measured as the average cost shares within the same plant size class in the five-digit industry in a given year. In calculating various input elasticity, plants are grouped into three size classes according to the number of employees: 50 or less, 51-300, and over 300. Thus, factor elasticity of plants was permitted to vary across industries and size classes, as well as over time.

REFERENCES

Ahn, S. (2001), "Firm dynamics and productivity growth: a review of micro evidence from OECD countries", *OECD Economics Department Working Paper 297,* OECD Economics Department.

Ahn, S., Fukao, K. and Kwon, H.U. (2005), "The internationalization and performance of Korean and Japanese firms: an empirical analysis based on micro-data", *Discussion Paper 05008*, Research Institute of Economy, Trade and Industry (RIETI).

Aw, B.Y., Beeyan and Rodrik, D. (1994), "Distribution Politics and Economic Growth" *The Quarterly Journal of Economics,* Vol. 109, pp. 465~490.

Aw, B.Y., Chen, X. and Roberts, M.J. (2001), "Firm-level evidence on productivity differentials and turnover in Taiwanese manufacturing", *Journal of Development Economics,* Vol. 66, pp. 51-86.

Baldwin, J.R. and Gu, W. (2006), "Plant turnover and productivity growth in Canadian manufacturing, *Industrial and Corporate Change",* 15, 417-465.

Barrett, G. and Donald, S. (2003), "Consistent Tests for Stochastic Dominance", *Econometrica* 71, 1, pp. 71-104.

Chernozhukov, V. (2002), "Inference on Quantile Regression Process, an Alternative", Working Paper, MIT.

Disney, R., Haskel, J. and Heden, Y. (2003), "Restructuring and productivity growth in UK manufacturing", *The Economic Journal 113*, pp. 666-694.

Foster, Lucia, Haltiwanger, and Krizan, C.J. (2001), "Aggregate Productivity Growth: Lessons from Microeconomic Evidence", New Developments in Productivity Analysis, *Studies in Income and Wealth,* Vol. 63, University of Chicago Press.

Fukao, K., Kim, Y.G. and Kwon, H.U. (2006), "Plant turnover and TFP dynamics in Japanese manufacturing", *Hi-Stat Discussion Paper Series, d06-180*, Institute of Economic Research, Hitotsubashi University.

Geroski, P.A. (1995), "What Do We Know about Entry?", *International Journal of Industrial Organization,* Vol. 13, No. 4, pp.421-440.

Good, D.H. (1985), "The Effect of Deregulation on the Productive Efficiency and Cost Structure of the Airline Industry", *Ph.D. dissertation,* University of Pennsylvania.

Grossman, Gene, M. and Helpman, E. (1991), *"Innovation and Growth in the Global Economy*", The MIT Press.

Hahn, C. H. (2000), "Entry, Exit, and Aggregate Productivity Growth: Micro Evidence on Korean Manufacturing" *OECD Economics Department Working Papers No. 272.*

Hahn, C. H. (2003), *"Entry, Exit, and Aggregate Productivity Growth: Evidence on Korean Manufacturing"*, Korea Development Institute Policy Study 2003-2002.

Haltiwanger, J. (2000), "Aggregate Growth: What Have We Learned from Microeconomic Evidence?", *OECD Economics Department Working Paper No. 267,* OECD, Paris.

IMF (2005), "Republic of Korea: 2004 article IV consultation-staff report; staff statement; and public information notice on the executive board discussion", *IMF Country Report 05/49,* International Monetary Fund, Washington, D.C.

Kang, D.S. (2005), "Corporate Distress and Restructuring Policy of Korean Small and Medium-sized Enterprises: Role of Credit Guarantee Scheme", *Paper presented at the 2005 KDI-KAEA conference,* Korea Development Institute.

Kim, J. H. (2005), "The Role of SMEs in the Innovation-driven Economy", *Korea Development Institute Research Report 2005-2005.*

Koo, J. and Kiser, S.L.(2001), "Recovery from a financial crisis: the case of South Korea", *Economic and Financial Review,* Q IV, 24-36.

Linton, O., Maasoumi, E. and Whang Y. J.(2005), "Consistent Testing for Stochastic Dominance under General Sampling Schemes", Cowles Foundation, SMU, and the LSE EM/02/433, forthcoming, REVIEW OF ECONOMIC STUDIES.

Maasoumi E. and Heshmati, A. (2005), "Evaluating Dominance Ranking of PSID Incomes by Various Household Attributes", in Gianni Betti and Achille Lemmi, Eds., *Advances on income inequality and concentration measures: Collected papers in memory of Corrado Gini and Max O. Lorenz*, forthcoming, and IZA Discussion Paper No. 1727.

McFadden, D. (1989), "Testing for stochastic dominance", in Part II of T. Fomby and T.K. Seo (eds.) Studies in the Economics of Uncertainty (in honor of J. Hadar), Springer-Verlag.

OECD (1998), *"Science, Technology and Industry Outlook, OECD"*, Paris.

OECD (2003), "*The Sources of Economic Growth in the OECD Countries",* OECD.

Oh, I., Heshmati, A., Baek, C. and Lee, J-D (2006), "Comparative Analysis of Firm Dynamics by Size: Korean Manufacturing", RATIO Working Papers No. 94, The RATIO Institute, Sweden.

Stokey, N. L. (1988), "Learning by doing and the Introduction of New Goods", *Quarterly Journal of Economics,* Vol.96, No 4, pp.701-717.

INDEX

A

B

C

D

E

F

J

K

L

M

N

O

P

T

U

V

W